A Fifty-Year Love Affair
with Organic Chemistry

A Fifty-Year Love Affair
with Organic Chemistry

William S. Johnson, 1913–
III

PROFILES, PATHWAYS, AND DREAMS
Autobiographies of Eminent Chemists

Jeffrey I. Seeman, Series Editor

American Chemical Society, Washington, DC

Library of Congress Cataloging-in-Publication Data

Johnson, William S., 1913–1995
    A fifty-year love affair with organic chemistry / William S. Johnson

      p.   cm. — (Profiles, pathways, and dreams, ISSN 1047–8329)
    Includes bibliographical references and index.

    ISBN 0–8412–1834–X (alk. paper)

      1. Johnson, William S., 1913–1995. 2. Chemists—
    United States—Biography. 3. Chemistry, Organic—United
    States—History—20th century.

      I. Title. II. Series.
    QD22.J58A3   1997
      540'.92—dc21   97–15695
    [B]   CIP

Jeffrey I. Seeman, Series Editor

The paper used in this publication meets the minimum requirements of American National Standard for Information Sciences—Permanence of Paper for Printed Library Materials, ANSI Z39.48–1984.   ∞

PRINTED IN THE UNITED STATES OF AMERICA

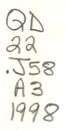

Foreword

In 1986, the ACS Books Department accepted for publication a collection of autobiographies of organic chemists, to be published in a single volume. However, the authors were much more prolific than the project's editor, Jeffrey I. Seeman, had anticipated, and under his guidance and encouragement, the project took on a life of its own. The original volume evolved into 22 volumes, and the first volume of Profiles, Pathways, and Dreams: Autobiographies of Eminent Chemists was published in 1990. Unlike the original volume, the series was structured to include chemical scientists in all specialties, not just organic chemistry. Our hope is that those who know the authors will be confirmed in their admiration for them, and that those who do not know them will find these eminent scientists a source of inspiration and encouragement, not only in any scientific endeavors, but also in life.

Contributors

We thank the following corporations and Herchel Smith for their generous financial support of the series Profiles, Pathways, and Dreams.

Akzo nv

Bachem Inc.

DuPont

Duphar B.V.

Eisai Co., Ltd.

Fujisawa Pharmaceutical
    Co., Ltd.

Hoechst Celanese Corporation

Imperial Chemical Industries PLC

Kao Corporation

Mitsui Petrochemical Industries,
    Ltd.

The NutraSweet Company

Organon International B.V.

Pergamon Press PLC

Pfizer Inc.

Philip Morris

Quest International

Sandoz Pharmaceuticals
    Corporation

Sankyo Company, Ltd.

Schering–Plough Corporation

Shionogi Research Laboratories,
    Shionogi & Co., Ltd.

Herchel Smith

Suntory Institute for Bioorganic
    Research

Takasago International
    Corporation

Takeda Chemical Industries, Ltd.

Unilever Research U.S., Inc.

Profiles, Pathways, and Dreams

Titles in This Series

About the Editor

JEFFREY I. SEEMAN received his B.S. with high honors in 1967 from the Stevens Institute of Technology in Hoboken, New Jersey, and his Ph.D. in organic chemistry in 1971 from the University of California, Berkeley. Following a two-year staff fellowship at the Laboratory of Chemical Physics of the National Institutes of Health in Bethesda, Maryland, he joined the Philip Morris Research Center in Richmond, Virginia. In 1983–1984, he enjoyed a sabbatical year at the Dyson Perrins Laboratory in Oxford, England, and claims to have visited more than 90% of the castles in England, Wales, and Scotland.

Seeman's 100 published papers, patents, and books include research in the areas of photochemistry, nicotine and tobacco alkaloid chemistry and synthesis, conformational analysis, pyrolysis chemistry, organotransition metal chemistry, the use of cyclodextrins for chiral recognition, and structure–activity relationships in olfaction. He was a plenary lecturer at the Eighth IUPAC Conference on Physical Organic Chemistry and has been an invited lecturer at numerous scientific meetings and universities. From 1989 to 1994, Seeman served on the Petroleum Research Fund Advisory Board, and he currently serves on the advisory board of *The Journal of Organic Chemistry*. *Seeman continues to count Nero Wolfe and Archie Goodwin among his best friends.*

Dedication

Dedicated to my co-workers, my wife Barbara, and my secretary Carolyn Southern.

Contents

Photographs

Preface

"HOW DID YOU GET THE IDEA—and the good fortune—to convince 22 world-famous chemists to write their autobiographies?" This question has been asked of me, in these or similar words, frequently over the past several years. I hope to explain in this preface how the project came about, how the contributors were chosen, what the editorial ground rules were, what was the editorial context in which these scientists wrote their stories, and the answers to related issues. Furthermore, several authors specifically requested that the project's boundary conditions be known.

As I was preparing an article[1] for *Chemical Reviews* on the Curtin–Hammett principle, I became interested in the people who did the work and the human side of the scientific developments. I am a chemist, and I also have a deep appreciation of history, especially in the sense of individual accomplishments. Readers' responses to the historical section of that review encouraged me to take an active interest in the history of chemistry. The concept for Profiles, Pathways, and Dreams resulted from that interest.

My goal for Profiles was to document the development of modern organic chemistry by having individual chemists discuss their roles in this development. Authors were not chosen to represent my choice of the world's "best" organic chemists, as one might choose the "baseball all-star team of the century". Such an attempt would be foolish: Even the selection committees for the Nobel prizes do not make their decisions on such a premise.

xvii

The selection criteria were numerous. Each individual had to have made seminal contributions to organic chemistry over a multidecade career. (The average age of the authors is over 70!) Profiles would represent scientists born and professionally productive in different countries. (Chemistry in 13 countries is detailed.) Taken together, these individuals were to have conducted research in nearly all subspecialties of organic chemistry. Invitations to contribute were based on solicited advice and on recommendations of chemists from five continents, including nearly all of the contributors. The final assemblage was selected entirely and exclusively by me. Not all who were invited chose to participate, and not all who should have been invited could be asked.

A very detailed four-page document was sent to the contributors, in which they were informed that the objectives of the series were

1. to delineate the overall scientific development of organic chemistry during the past 30–40 years, a period during which this field has dramatically changed and matured;

2. to describe the development of specific areas of organic chemistry; to highlight the crucial discoveries and to examine the impact they have had on the continuing development in the field;

3. to focus attention on the research of some of the seminal contributors to organic chemistry; to indicate how their research programs progressed over a 20–40-year period; and

4. to provide a documented source for individuals interested in the hows and whys of the development of modern organic chemistry.

One noted scientist explained his refusal to contribute a volume by saying, in part, that "it is extraordinarily difficult to write in good taste about oneself. Only if one can manage a humorous and light touch does it come off well. Naturally, I would like to place my work in what I consider its true scientific perspective, but . . ."

Each autobiography reflects the author's science, his lifestyle, and the style of his research. Naturally, the volumes are not uniform, although each author attempted to follow the guidelines. "To write in good taste" was not an objective of the series. On the contrary, the authors were specifically requested not to write a review article of their field, but to detail their own research accomplishments. To the extent that this instruction was followed and the result is not "in good taste", then these are criticisms that I, as editor, must bear, not the writer.

As in any project, I have a few regrets. It is truly sad that Egbert Havinga and Herman Mark, who each wrote a volume, and David Ginsburg, who translated another, died during the course of this project. There have been many rewards, some of which are documented in my personal account of this project, entitled "Extracting the Essence: Adventures of an Editor" published in *CHEMTECH*.[2]

Acknowledgments

I join the entire scientific community in offering each author unbounded thanks. I thank their families and their secretaries for their contributions. Furthermore, I thank numerous chemists for reading and reviewing the autobiographies, for lending photographs, for sharing information, and for providing each of the authors and me the encouragement to proceed in a project that was far more costly in time and energy than any of us had anticipated.

I thank my employer, Philip Morris USA, and J. Charles, R. N. Ferguson, K.Houghton, H. Grubbs, and W. F. Kuhn, for without their support Profiles, Pathways, and Dreams could not have been. I thank the staff of ACS Books for their hard work, dedication, and support. Each reader no doubt joins me in thanking 24 corporations and Herchel Smith for financial support for the project.

I thank my children, Jonathan and Brooke, for their patience and understanding; remarkably, I have been working on Profiles for more than half of their lives—probably the only half that they can remember! Finally, I again thank all those mentioned and especially my family, friends, colleagues, and the 22 authors for allowing me to share this experience with them.

JEFFREY I. SEEMAN
Philip Morris Research Center
Richmond, VA 23234

April 19, 1993

[1] Seeman, J. I. *Chem. Rev.* **1983**, *83*, 83–134.
[2] Seeman, J. I. *CHEMTECH* **1990**, 20(2), 86–90.

Editor's Notes

For more than a week in the early summer of 1985, Bill Johnson considered the invitation to join Profiles, only to call with a negative response. "I'm sorry to turn you down," he said, "but there is no way that I could tell the results of all of my students. It would not be fair to leave any of them out."

Not all who were asked to contribute to the series agreed to do so. The reasons were all unique, but Johnson's was distinctive in a very special way. He was primarily concerned about the feelings of his former students and collaborators.

Fortunately, Johnson changed his mind. His passing on August 19, 1995 makes this volume even more special, and his words ring out as I reread them. I know that Bill made every effort to include as many people and as large a variety of chemistry as possible.

People and chemistry had both been the two top priorities for Johnson, with the former taking precedence. To my knowledge, no contributor to Profiles wrote to as many of his former students for updated information and sent his draft manuscript to so many individuals. Even Mary Fieser received a copy with Johnson's request for the Amherst–Williams football game picture (*see* page 17). "I was in the midst of reading page proofs on *Reagents* when the mail arrived with your autobiography," Mary Fieser wrote Johnson. "So I thought it would be a pleasant change to glance at that. Instead I was so fascinated that I spent the rest of the day reading the text and then the formulas. Today I have [been] rereading the text and formulas together." Mrs. Fieser then offered two pages of suggestions and comments, to which Johnson responded, "On reading your letter, Barbara [Johnson's wife] noted that your suggested changes in the manuscript were mostly ones which would make me look like a nicer person. I believe Barbara is right, and even though you may not have realized it, your letter is a dear, sweet gesture besides being very helpful."

When I informed Bill that there would be an introductory essay, this Editor's Note, for his contribution, he responded with a very hearty laugh and he chuckled, "Tell them there isn't an ounce of truth in what they are going to read in the following!"

In fact, Johnson was firmly committed to the truth. In the Spring of 1987, when *C&E News* announced the first annual W. S. Johnson Symposium in Organic Chemistry, the headline read "Stanford University honors former professor." Johnson responded in a letter to the Editor of *Chemical & Engineering News*: "At first I was inclined to regard [this statement] as innocuous. Gradually, however, it became apparent that many people felt I had given up Chemistry...I would like to assure my friends and future collaborators that I am still in business."

It is very fitting that Stanford honor Johnson with these annual symposia, the 13th will be held in 1998. According to Carl Djerassi, Johnson "creat[ed] the modern chemistry department at Stanford." One of Johnson's first decisions regarding his Profiles book was that he would tell the story of his Chairmanship at Stanford and the hiring of Djerassi, Flory, Taube, van Tamelen, McConnell, and others. Johnson was very proud of the Stanford chemistry tradition.

Lest one conclude that Johnson is an 'old softy,' remember that one does not achieve without commensurate hard work. "I put the pressure on my students also. I always worked Saturday and Sunday mornings, and they felt a little guilty if they did not come in then also. Actually, for many years, we had our group meetings on Saturdays. I never 'made-the-rounds' to check on who was in. However, I did drop in to talk with my students...I always felt that they might want to feel that *I've seen them*! Remember, they once called me 'Black Bill'!" Consistent with this nickname, one of Bill's former students summarized that he "could go on about Bill's personal/professional attributes for days, but, in brief, they are intense focus, unfailing optimism, robust humor, IRON WILL."

Johnson's book moves along nicely, smoothly and interestingly until suddenly the reader is jerked, like a dog on a snapped leash. Without warning, Johnson tells a story of violence, an event that occurred in Johnson's laboratory at The University of Wisconsin nearly 50 years ago. The reader is moved from chemistry to human frailty and suffering. Was Johnson still haunted by these very old memories, still trying to set the record straight? He explains that "grossly distorted versions have been circulating even to this day." Such a story does have an important place in an autobiography of a famous scientist. Chemistry is performed and appreciated by people, not divorced from the everyday world of reality. "[Many] who have read the manuscript seem to feel that the...story should be told, perhaps because it was an event which had a significant effect on me and a number of my students, drawing us together in a lasting way."

Perhaps the greatest compliment that can be paid to anyone is peer recognition. In his 76th year, Johnson received the Arthur C. Cope Award. "Bill was eager to give credit to his student's efforts," recalls a former student, who shared Bill's letter to him. Johnson wrote, on December 14, 1988, "Many thanks for the congratulatory letter. I was truly surprised but very pleased about the Cope Award selection, particularly because of its reflection on my former collaborators like yourself."

The Cope award's purpose is "To recognize outstanding achievement in the field of organic chemistry, the significance of which has become apparent within the [past] five years..." Johnson found significant rate and yield enhancements of polyene cyclizations by the novel expedient of positioning cation-stabilizing groups that stabilize the transition state(s) of the process. These observations suggested an attractive mechanism for the action of oxidosqualene cyclases. Among the previous recipients are such people as R. B. Woodward, R. Hoffmann, D. J. Cram, E. J. Corey, A. Eschenmoser, and Johnson's special friend Gilbert Stork. It is a wonderful statement to be so 'senior' in age and so 'young' in achievement. Johnson's receipt of this award brought many of his friends much personal satisfaction and sunshine.

Coda

I bring to your attention the epilogue which begins on page 193. "A Perspective on the Final Chemistry from Johnson's group" was written by three of Bill Johnson's longest and closest friends along with one of his students. This essay compliments Bill's own words and allows his final chemistry to be within the pages of his book. Most important, Paul Bartlett, Ted Bartlett, Jack Roberts, and Gilbert Stork have added a warm, very personal final touch which, sadly, Bill himself could not provide. I take the liberty of repeating the final sentence of this epilogue. "All who knew him will forever admire his chemical style and personal philosophies."

Coda to a Coda

Madeline Jacobs, Editor of *Chemical and Engineering News*, reported in the January 12, 1998 issue of that substantive journal, "Over the course of three months in 1997, we asked *C&EN* readers to nominate their choices for *C&EN*'s 'Top 75 Distinguished Contributors to the Chemical Enterprise' during the 75 years of *C&EN*'s existence...Readers have come up with a superlative group of contributors, representing the diversity within the far-flung chemical enterprise..." William Summer Johnson was one of "*C&EN*'s Top 75." This is an honor that Bill most certainly would have enjoyed, particularly because he was joined by so many of his closest friends, including Carl Djerassi, Jack Roberts, and Gilbert Stork. I understand that Barbara Johnson plans to attend the next ACS National Meeting in Boston, where this recognition will be celebrated. I look forward to seeing her there, to celebrate Bill's life and legacy.

The words "coda," "coda to a coda," and "coda to a coda to a coda" were introduced to me by Derek Barton, who, upon multiple requests, added three codas to his fine autobiography, *Some Recollections of Gap Jumping*. Very sadly, since the writing of the final version of this Editor's Note, both Derek and Vladimir Prelog have passed away. The passing of these great scientists and fine human beings is a sadness for all of us

Special Acknowledgements to...

Ted Bartlett (Fort Lewis College) and *Ray Conrow* (Alcon Laboratories), former students, colleagues, and friends of Bill Johnson, who fully participated in the review of the final manuscript, the galleys, and the page proofs. In addition, Ted provided all the chemical structures in computerized format. It is difficult to complete the publication of a book after the passing of the author, and the enthusiastic participation of Ted and Ray ensured the quality that Bill's autobiography deserved.

Ted Bartlett, Paul Bartlett (University of California, Berkeley), *John D. Roberts* (CalTech), and *Gilbert Stork* (Columbia University) for writing the Epilogue.

A Fifty-Year Love Affair
with Organic Chemistry

William S. Johnson

My Introduction to Chemistry

Born in New Rochelle, New York, in 1913, I was the second of three children and the black sheep of a family of scholars. As a youngster I was the one who got into trouble (not serious by today's standards). I had little interest in school. My predilections included listening to and playing jazz (piano and tenor sax), building radios and sound equipment, and dating girls. By the end of my sophomore year in the New Rochelle High School I had been so inattentive to studies, had cut classes frequently, and accumulated so many "detention" hours that my academic future looked very bleak. My father, after being told by my school principal that I would never amount to anything, decided to send me to a private school for boys: his alma mater, Governor Dummer Academy, founded in 1763 as a preparatory school for Harvard. My sister (two years my senior and an A student in high school) and mother both thought that this was a gross waste of money. However, my father kept insisting, "I think Bill has potential." Because it was not possible for me to pursue my hobbies at Dummer and I found the teachers stimulating, I did quite well in my classes. During that year (1930–1931) the Great Depression hit my family. My father, who should have been a teacher of English literature but ended up in the field of business journalism (author of the book, *The Sales Strategy of John H. Patterson*), lost nearly all of his worldly possessions including our house, and we moved in with friends. Faced with no means to return to Dummer, I suddenly realized that this was what I really wanted more than anything. The school was also in financial trouble but the headmaster, Ted Eames, managed to come up with a full scholarship for me. He was indeed my great benefactor and I suspect that some of these funds came from his own pocket.

3

Isabelle and W. S. Johnson, the latter with his pianola roll, New Rochelle, NY, ca. 1915.

In 1931–1932 I had the top grades in the school and won a scholarship to Ted Eames' alma mater, Amherst College, thanks to his strong backing. My mother and sister could never quite believe what was happening, especially when I became Phi Beta Kappa in my junior year at Amherst and graduated *magna cum laude*. This occurred in the wake of my sister making a reverse and earning a C average at Mount Holyoke College. My father, bless his heart, lived long enough to see me earn a doctorate in chemistry and obtain a teaching job.

Now I shall put on record what my life was like before going away to Dummer Academy. Unlike my sister Isabelle, who was often bored or depressed, I was a very happy child. My brother Tom, who was six years my junior, was also contented but was a serious scholar who always did well in school. Isabelle and I were very close friends and I would listen to her complain about life by the hour because she always seemed grateful for the atten-

Bill, 2–3 years old, 1915–1916, New Rochelle, NY.

tion I gave her even though I really could not empathize. I found my own life so full of interesting things to do that there never seemed to be enough time. I was always deeply absorbed in several things at once.

At the age of 13 I witnessed a demonstration of home radio reception, which was only starting to become available. The "magic" so enthralled me that for several years I spent all of my spare time building radio sets and trying to establish new records for bringing in distant stations, "DXing". My first set had a simple crystal cat-whisker detector without any amplification and would bring in signals (monitored by earphones) from only two nearby stations, WJZ (New York) and WOR (New Jersey). Upgrading with a varicoupler improved reception, but soon I graduated to tube sets that I constructed for friends as well as myself.

Bill and Isabelle with their father Roy, about 1917.

My prize was a multitube superheterodyne that would bring in long-wave signals from across the continent. These sets were constructed from schematic drawings, and sometimes a photograph of the assembled product was available. This involved cutting a baseboard of hardwood and an instrument panel of bakelite, which was mounted perpendicularly by brackets. Parts procured from radio shops in New York were mounted, and the wiring was generally done with heavy bus-bar which, where necessary, was insulated with spaghetti tubing. All connections were soldered. Turning on a newly constructed set for the *first* time was always an exciting, and sometimes thrilling experience. My friend, Wilder Pray, and I started a radio repair business that did not amount to much for lack of time.

I was equally absorbed in music—listening as well as performing. I started with the usual classical piano lessons, which were not inspiring for a 13 year old, but I kept them up partly because I had an unreciprocated crush on my teacher, Dorothy

Bill, Tom, mother Josephine, Isabelle.

Isabelle Johnson (Spencer), me, and Tom Johnson, in a photographer's studio, ca. 1920.

Morey, who was in her 20s. Then, I was more interested in jazz, which I studied on my own at the piano. I also studied banjo, which I mastered sufficiently well for my parents to buy me a fine Gibson four-string tenor banjo that I played in a dance band called the High Hats. My real love was the tenor saxophone, which I took quite seriously. Also I played bass viol (classical music) in orchestras at prep school and college.

During the summer when I was 12 years old, my father Roy, whose company I always enjoyed immensely, took me on a most memorable vacation involving a very rough, overnight steamboat trip from New York to Boston during a violent north-easter storm. The next day involved some fun at an amusement park and then I witnessed my first professional baseball game, at

Bill in his early teens.

Fenway Park. That evening we boarded an overnight side-wheeler steamboat, which took us peacefully from Boston to Portland, Maine, where we took a train to Shelburne, New Hampshire. We were met by Miss Gates' handyman in a horse and carriage, which delivered us to Gates Cottage. My father introduced me to the White Mountains where we hiked over the same trails that my father as a boy had traversed, and in one case blazed, with his father. Like my father, I immediately fell in love with hiking in this area as well as the nearby Presidential Range. I have returned to this area frequently, often accompanied by my wife Barbara and sometimes by our friends.

The Johnson family, ca. 1923, at Gates Cottage in the White Mountains, Shelburne, NH, one of my favorite haunts, where we often vacationed. Top, left to right: grandfather, Thomas Lyn Johnson; mother, Josephine Summer; brother, Thomas Lyn, II; father, Roy Wilder; bottom, sister, Isabelle Wilder; and myself, August 1923.

My father Roy as I first remember him, ca. 1917.

My father was a born teacher and I learned much from him during our long hours together. When he was a freshman at Harvard he roomed with Clarence Lewis, who became a lifelong friend even though my father dropped out of school at the end of his freshman year because he found the classes boring. Lewis, however, went on to eventually become a professor of philosophy at Harvard where he made a worldwide reputation for his concept of symbolic logic, which is to this day regarded as an important tool in the field. I had some contact with the Lewis family when I was in graduate school at Harvard and later on at Stanford, where he had accepted an appointment.

Many years after my father's death at age 60, "Uncle" Clarence told me that he regarded my father as one of the most intelligent people he had ever known. He told how instead of going to class, Roy would amuse himself by taking an alarm clock apart and reconstructing it so as to resemble a creation that would actually move across the room under its own power. He also told how Roy, on finding the examinations unchallenging, would "play horse" with the questions, which generally annoyed the teacher into giving him a poor grade.

Thus my father was highly educated by self-teaching. He read avidly and was a master of several languages including Latin and Greek. On one occasion when he was in high school the teacher assigned a paragraph from Shakespeare's *The Tragedy of Julius Caesar* to be memorized. My father, unknown to others, had by then read all of Shakespeare's works, and was called on to recite. Not having paid any attention to the assignment, he did not have any idea where to begin. So he started at the beginning of the play and recited all of Act I, Scene I, as the teacher, in amazement, urged him on. He could have done the whole play but time ran out. Dad hardly ever flaunted this talent because he did not believe that total recall was a form of real intelligence, which he defined as "imagination and the ability to use it", as in all forms of artistry and science.

Because the class of 1936 at Amherst was coping with the depth of the Depression, most of us felt very lucky just to be in school and therefore became relatively serious scholars. I was totally self-supported, but managed well enough with a tuition fellowship and various jobs. I tended furnace and did house as well as yard work at the Emily Dickinson house (then privately owned by the Parke family) for the privilege of rooming there. For my meals I washed dishes at the Delta Upsilon fraternity and made a few dollars for incidentals by playing saxophone in dance bands. Subsistence during the summers was a challenge. In 1933 I played sax with a dance band at a hotel in the Catskills and received room and board plus $5 a week. In 1934 I made my first trip to Europe by playing for dances on the ocean liners. In 1935 I stayed in Amherst without work, except for an occasional dance job, ate poorly, and ended up with pneumonia, a serious disease in those preantibiotic days.

Returning to the summer of 1933, I was involved with some very talented professional musicians, some of whom eventually played with name bands. Performing with these people was such an exciting experience that I seriously considered trying to make it as a professional.

However, when I returned to Amherst and had my first exposure to organic chemistry, I became hooked by reading James B. Conant's textbook. Even this early I was particularly fascinated by the powerful concepts of molecular architecture that provided the potential for determining the structure of and then constructing complex molecules from readily available commercial chemicals. After that there was no doubt in my mind about what I wanted to do. Naturally I majored in chemistry, and during my senior year I was invited to remain at Amherst an additional year as instructor to teach organic chemistry in place of Robert Whitney, who was taking sabbatical leave. With this appointment on my resume, I was able to land a job for the summers of 1936–1939 at the Eastman Kodak Company Research Laboratories, where I got a lot of practical experience at the bench in the sensitizer dye department of Leslie Brooker with whom, as it turned out, I enjoyed a life-long friendship.

My teaching and industrial experience, along with the backing of Ralph Beebe and Bob Whitney, helped me win a full-time fellowship at Harvard, which I chose with the aim of working for Professor E. P. Kohler. When I saw him, he said "I'm all filled up, go see Fieser." With considerable luck and a highly supportive mentor in Louis Fieser, I completed work for the doctoral degree in 22 months of residence: September 1937 to January 1940 (summers at Kodak). (I think Fieser promoted this pace so that he could brag of having the student who obtained the doctorate in record time.) My project involved the synthesis of a variety of

John X, drums; Harold Raby, trumpet, Eddie Ciruti, piano; Bill Johnson, tenor sax; Bill Kirdy, alto sax. The Cunard Line wanted a "college" dance band; however, half of the group were professionals put together at the "last minute" so the music would be acceptable. The photograph is in the summer of 1933 on the Transylvania while en route from New York to Scotland. Later on, during the war, this boat was converted to a troop transport and was sunk by a German submarine.

substituted benzanthracenes and chrysenes for screening as carcinogens.

Louis Fieser was an excellent major professor for me. I received lots of attention but at the same time he encouraged me to pursue my own ideas. We had three publications and he invited me to participate in the preparation of the manuscripts, during which I learned a great deal about writing from a real master. His book entitled *Natural Products Related to Phenanthrene* (1936) contained the foremost treatise on steroids, which along with a graduate course he gave sparked my interest in the field. At that time the first steroid total synthesis was in progress in the laboratory of Werner Bachmann at Michigan, and when I later learned about it I was greatly inspired.

Louis was a man of great vitality and physical strength. On one occasion when I was working in the lab alone, he stopped by just as my nitrogen tank was running low in the middle of a critical experiment that needed my attention. He immediately dashed out of the lab and soon returned carrying a fresh, *full-sized* tank of nitrogen on his back. Louis had attended Williams College and played varsity football there. Because Williams and Amherst were archrivals, he and I made a bet of 25¢ on the big game in 1939 and the bet was an annual event for more than 30 years.

The last record I have on this matter is a letter dated December 14, 1971, in which Louis wrote: "Thanks for coming across with the 1971 wager. I note that my school now admits girls, who can provide some highly expedited cheer leading, so your Amherst boys had better watch out." As a member of the Surgeon General's committee that produced the original study on the relation of smoking to health, Fieser was convinced of the cancer hazards yet he continued to be a chain cigarette smoker. When I asked him why he took this risk, he replied that he had a very effective metabolism and felt he could detoxify the carcinogens. In 1963 he had surgery for lung cancer and never regained his full vitality, although he was still active and started the important *Reagents for Organic Synthesis* series, which Mary Fieser has continued since his death in 1977 when he was 78 years old.

Another individual who was very important to me was Fieser's senior postdoctoral student, E. B. Hershberg. The first step of my Ph.D. research project was to repeat the known high-

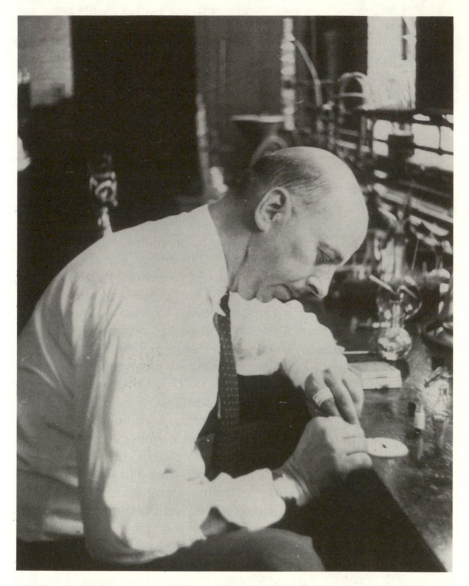

Louis Fieser, as we students would often find him, about 1950.
(Photo courtesy of Harvard University Archives.)

Photo taken by me of Louis Fieser, Barbara, and Mary Fieser at the Amherst-Williams football game in 1954. (Photo courtesy of Mary Fieser.)

pressure hydrogenation of phenanthrene over copper chromium oxide to produce 9,10-dihydrophenanthrene. With the typical disregard most academic institutions had for safety practices, I was directed to work with a bomb at booster-pump pressure of hydrogen without any shield in an open laboratory where others also worked at the bench. Fieser gave me a large supply of beautifully crystalline phenanthrene indicating that it should be suitable for hydrogenation. Numerous attempts to hydrogenate the material, even after purification by recrystallization, distillation, and chromatography, failed completely. After about 6 weeks I was a desperate young man with hands blistered from tightening the bolts on the bomb. Then E. B. came to my rescue and suggested that just as benzene from coal tar is contaminated by thiophene, so phenanthrene could well be similarly contaminated with sulfur analogs, which would poison the catalyst. E. B. ran a sodium-fusion test for sulfur on my phenanthrene sample and it was strongly positive.

    The technique in those days for desulfurization was to

heat and stir the substrate with a large excess of molten sodium (Raney nickel desulfurization was not yet an established art). At last the hydrogenation of material thus treated worked like a charm and I felt that I might be able to get a Ph.D. after all. The workup procedure on a sizeable scale was hazardous to say the least and I enjoyed a few fires. At that time, fires were rather common in organic laboratories. I recall seeing Fieser swirling an open Erlenmeyer containing pentane and methanol for a recrystallization while warming it over a free flame; if it caught on fire he nonchalantly smothered it with a towel. He said that using hot plates was too slow.

    Many of us learned a lot from E. B., who was a superb experimentalist as well as a fine, generous person. He taught us never to believe what is on the label of a chemical, even from a "reliable" commercial vendor, and to always confirm the identity and purity by some physical measurements. A favorite remark of his when we ran into troubles with questionable reagents was "never assume anything". E. B. also taught us good laboratory manners. He said "if you borrow some equipment from a colleague, always return it just as *soon* as you are through with it and be sure that it is cleaner, if possible, than it was when you borrowed it." His comment in regard to my original specimen of phenanthrene was, "Beauty is no criterion of purity."

    From January to August 1940 I served as a postdoctoral assistant to Professor Reginald P. Linstead at Harvard while I was waiting to assume a teaching job at Wisconsin. Linstead had just broken his leg while skiing (his first time) with George Kistiakowsky; hence, I had little contact with my mentor who on several occasions shook his finger admonishingly at me saying, "Johnson, never get on a pair of skis!" In fact I took his advice, and have often told this story to my colleagues, most of whom, such as Carl Djerassi and Doug Skoog, did not take it seriously, to their loss.

    In September 1940, I moved to the University of Wisconsin as instructor, a job that I surely would not have gotten without Louis Fieser's strong backing. Thus 1940 was a very eventful year considering that I also met Barbara Allen in Cambridge. Soon after I moved to Wisconsin, she joined me in December and we were married on the 27th. My salary was $2000/year and our first apartment cost $33/month. It was on the fourth floor (no elevator) and we shared a bathroom with two other apartments, one of which was occupied by a prostitute. My marriage was most fortunate for me in many respects. As to my professional life, it turned out that Barbara, although a literature major, vicariously shared my enthusiasm and often accompanied me to the lab in the evenings or on weekends, bringing along her own work (writing) or reading. It is hard to overemphasize how important her support has been to my career.

Barbara helped me in a variety of ways ranging from proofreading papers to planning and serving as hostess for annual parties for my research groups that were held at Thanksgiving and Christmas time at Wisconsin, and on the Fourth of July around our swimming pool at Stanford. In this way she got to

Barbara Allen, before our marriage, at her home in Wilmington, Massachusetts, 1940.

know nearly all of my co-workers, who became very fond of her. Barbara was always ready to entertain academic and industrial visitors with whom I had professional interaction. Because we had no children she often traveled with me when I was invited to foreign countries. Her presence on these occasions gave me a feeling of delight and pride.

I was lucky to find myself in one of the best places for organic chemistry at that time. S. M. ("Mac") McElvain and Homer Adkins were top people in the field, and according to a basic rule, they attracted top students. Their "recent" students included such people as Art Cope and Karl Folkers. The quality of the organic students attracted to Wisconsin in the early 1940s was remarkable; I have never since seen a better collection of talent in any university. In those days Illinois, under the guidance of Roger Adams and Carl S. (Speed) Marvel, was comparably attractive to organic students, and for many years the two schools directed their undergraduate chemistry majors to each other for graduate training. I share C. David (Dave) Gutsche's (Ph.D. 1947) feelings about the Wisconsin scene in the 1940s as described in a letter he wrote me on October 27, 1986:

> The days at Wisconsin were heady ones for me, for it was my first exposure to the excitement of organic chemistry. It was so stimulating to be a part of a project whose leader was filled with such enthusiasm and vitality and to be in a department with fellow students like Gilbert Stork, Carl Djerassi, Jack Peterson, and Harvey Posvic, to mention just a few. I didn't realize at the time how special this group of people was, assuming it to be nothing more than a bunch of bright students typical of any good graduate school. But, it *was* a special group, and beyond any question, it was an experience that shaped the rest of my life and one in which you played the pivotal role—for which I am eternally grateful. My own career has been one of only modest achievement, but what success I have had can be credited directly to the superb start that I got in the Department of Chemistry at Wisconsin.

Bill being athletic at a pool party at his home, being held by Leo Bunes, ca. 1979. "We were doing acrobatics at one of our 4th of July swimming–picnic parties that Barbara and I gave for about 20 years." (Photo courtesy of R. Schmid and Dr. and Mrs. R. K. Muller.)

The organic group at Wisconsin in 1940: (left to right) Sam McElvain, W. S. Johnson, Homer Adkins, Al Wilds, and Mike Klein.

Well, those were heady days for me also as I was closely associated with my colleagues, particularly Mac, Homer, Al Wilds, (and later in the 1950s) Harlan Goering, and Gene van Tamelen, as well as with many of the students. Many of these contacts have developed into rich life-long personal as well as professional associations. Gene is still one of my colleagues at Stanford, and we have also kept in particularly close touch with the Goerings, whom we have joined on several occasions for vacations in the Caribbean. Harlan has always been one of my favorite consultants on physical organic issues; also he is a jazz and hi-fi enthusiast, which has brought us together on numerous occasions. The organic faculty at Wisconsin enjoyed a remarkable *esprit de corps*, and the operational affairs were conducted most amicably at our weekly luncheon meetings. The tone was set by the magnanimous attitude of the senior staff: Homer, Mac, and Mike Klein, who for example always volunteered for extra teaching duties to give the younger colleagues additional time for research. This giving attitude was contagious.

We young members of the faculty received all sorts of helpful tips from our elders. For example, on one occasion I was bragging to Mac about my student Dave Gutsche who was, among other things, completing his Ph.D. thesis work in a record time of two years and five months. Mac then said to me, "If you are lucky enough to attract to your research group several students of Gutsche's caliber, these people are likely to make you famous." That provocative remark permanently oriented my attitude regarding the matter of where the credit belongs.

With Dave Gutsche at his graduation ceremonies to receive the Ph.D. degree (Madison, 1947).

Speaking of Dave Gutsche, when he asked if I would attend the graduation ceremonies when he was to receive his Ph.D., I told him that I was not sure whether or not to accept his invitation. Actually I had considered such things as being a waste of time and had not gone to the ceremonies when I was awarded the Masters and Ph.D. at Harvard. Such behavior was generally acceptable because these affairs were usually overcrowded.

On the other hand, Dave, who was one of my students, was enthusiastic about the idea; moreover, I later learned that his family was coming to the affair. So I agreed to participate. I was so late in renting a gown that hoods for Harvard Ph.D.s were not available and the closest thing in appearance was an Ohio State hood for Master's degree, which I am wearing in the photo.

Now that I was "on my own" in 1940, I hoped to be able to satisfy my predilection for organic synthesis by doing something directed toward the total synthesis of steroids, which soon proved to be the hottest synthetic target of the time. Adolf Windaus, in his 1928 Nobel Lecture[1a] on steroids said, "The synthesis of such a substance appears to the chemist particularly difficult and up till now I have not dared to attempt it." By 1940 things had not changed significantly except for the classical synthesis of equilenin (*see* subsequent discussion).

In my last year at Harvard, I had heard Professor Werner Bachmann (Michigan) deliver a lecture on the first total synthesis of a steroid, namely equilenin (7, Scheme 1). This was the most significant multistep synthesis of its time—a true masterpiece. It set a standard of experimental virtuosity that has rarely been met and has never been exceeded even to the present day. The average yield per step from ketone 2 to *d,l*-7 was close to 90%. By today's standards, the strategy suffers from lack of stereoselectivity at the acrylate reduction step leading to 4; however, scientists were hardly concerned with this issue in 1940. In any case the Bachmann, Cole, and Wilds equilenin synthesis[1b] was a tremendous inspiration, and these men were always heroes to me.

I was delighted to find, on arriving at Wisconsin, that Bachmann's student Al Wilds had also just been hired as an instructor. Al and I had a very close relationship, particularly in the early days when we both were learning to become independent investigators. We tried out ideas on each other, exchanged

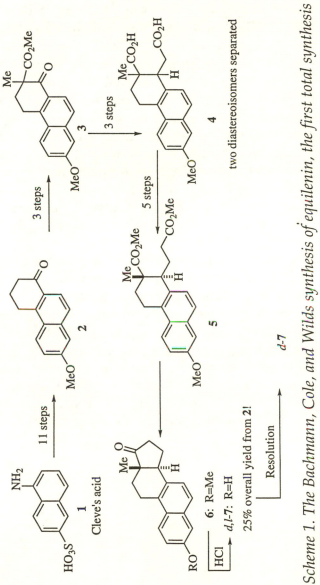

Scheme 1. The Bachmann, Cole, and Wilds synthesis of equilenin, the first total synthesis of a steroid, 1940.

thoughts about procedures, and read each other's attempts at writing papers. Eventually, we wrote a laboratory manual for beginning organic chemistry, (Johnson, W. S.; Klein, M. W.; Wilds, A. L. *Experiments in Organic Chemistry;* William C. Brown: Dubuque, IA, 1947, 1951). We devised most of the experiments for this book, and developed them in the laboratory ourselves. We both made excellent early progress in our research, much of it accomplished with our own hands, so that there was never any great concern about getting tenure. Indeed we both received promotions at the same time. Al and I were kindred spirits striving to emulate Bachmann's style of research. Al was more successful than I at this endeavor; however, his almost obsessive search for perfection has taken its toll. He has accumulated and written up a wealth of highly significant, meticulously performed research results that he has never felt were quite ready for publication. I am still trying to urge him to let the scientific public have the benefit of this fine work.

While we were still fairly young, Al and I were astonished, annoyed, and at the same time amused to discover that it was believed by many that we were archenemies. Since then I have come to realize that most chemistry departments are hotbeds of idle gossip, which often spreads all over the world. Be that as it may, conditioned by this early experience, I have never been inclined to believe any of the shocking stories I hear, unless they come from the "horse's mouth", such as Sir Robert Robinson's feelings about Sir Christopher Ingold (*see* subsequent discussion). Incidentally, Al and Carolyn Wilds are dear friends of ours to this day.

Early Days on My Own

Protocol for Synthesis Research in the 1940s

Imagine carrying out organic synthetic studies without the agency of IR, UV, NMR, MS, GC, or TLC, not to mention HPLC or X-ray crystallography. All we needed to do research was some glassware, rubber and cork stoppers, a cork-borer (ground-glass joints were a luxury then), some rubber tubing, a hot plate, a steam bath, a balance, and a water aspirator as well as a vacuum pump. The classical synthesis of equilenin was performed under these conditions, making this feat all the more remarkable. Our protocol, up through the early 1940s, was to purify products carefully by column chromatography and distillation or repeated recrystallization (melting point was widely used as a criterion of purity). Compositional combustion analyses on a "semi-micro" scale (35–100 mg samples were burned) were performed by our own hands on all new compounds, and this information was used as a criterion of purity and identity. Very crude molecular weight information could be obtained by the cryoscopic method, and in desperation a UV spectrum could be secured by using a Hilger quartz spectrograph. This second, laborious technique involved taking numerous photographs at various frequencies and plotting the "intensity" from the estimated density of the bands of developed photographic emulsion on glass plates. Fred Mathews, the first student of mine to obtain a Ph.D. (in 1943), learned and used this technique to obtain the UV spectrum of the Combe's cyclization product of the anil **8** (Scheme 2) resulting from condensation of acetylacetone with β-naphthylamine.[2] The UV data provided confirmatory evidence that the product was the linear benzoqui-

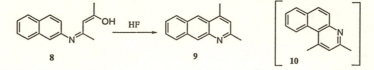

Scheme 2. The Combe's quinoline synthesis as applied to β-naphthylamine.

noline **9**, an unexpected result in view of the course of the Skraup reaction, which gives the angular isomer **10**.[2]

Some time thereafter we obtained one of the first Beckman DU spectrophotometers. Great excitement! One could produce a UV spectrum in less than a half hour by plotting intensity readings from a meter as you varied the frequency with a potentiometer. It was several years later that IR spectrometers were available to us as a routine facility, and this was a major breakthrough for obtaining structural information. Bob Woodward was one of the first to recognize the power of IR spectrometry, and with it he provided the major evidence for cracking the extraordinarily difficult problem of the β-lactam structure of penicillin.

Saul Winstein, barbecuing for Bob Woodward, at the Winsteins. (Photo courtesy of Sylvia Winstein.)

In this period we also had World War II to contend with. Some of us were luckily exempt from the draft because we were teaching organic chemistry to special premedical students who were in the Army ASTP and Navy V-12 programs. This, of course, was in addition to our regular teaching, and I can remember days when I had five lecture classes plus some laboratory duties.

The Stobbe Condensation

This reaction, the alkoxide-promoted condensation of aldehydes or ketones with succinic ester (*see* Scheme 3) intrigued me because of its synthetic potential for introducing a propionic side chain at the site of an aldehyde or ketone carbonyl group. This reaction eventually proved to be more efficient than the Reformatsky re-action followed by Arndt–Eistert homologation used in the aforementioned synthesis of equilenin. The classical procedure for the Stobbe condensation used diethyl succinate and sodium ethoxide in ethanol. The mixture became very dark, many by-products were formed, and in the benzophenone case (Scheme 3, R = Et) the oily half ester **11** was isolated in about 50% yield. A clue to the problem was the isolation also of some benzhydrol, evidently produced by a Meerwein–Ponndorff-type reduction of benzophenone by hydride anion from the ethoxide, which con-comitantly produced acetaldehyde, resulting in further messy condensations. Since that realization, I have always avoided the

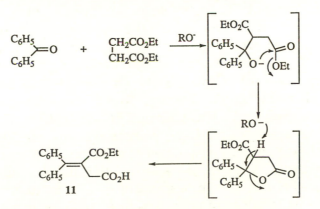

Scheme 3. The Stobbe condensation.

use of ethoxide anion for any sort of reaction that involves aldehydo or keto functions. Methoxide is a much weaker reducing
agent and works rather well in the Stobbe condensation; however,
t-butoxide is by far the best. With this reagent the "crude" reaction product (Scheme 3, R = *t*-Bu) was the beautifully colorless
crystalline half ester **11**, mp 120–124 °C, obtained in 90–94% yield.
This procedure is described in detail in a chapter on the Stobbe
condensation that Guido Daub and I eventually prepared for *Organic Reactions.*[3]

One of the plans I envisaged was utilization of the Stobbe
condensation with deoxybenzoin to produce, after hydrogenation
of the olefinic bond, a dicarboxylic acid which on double cyclization should give the 1,2-benzanthracene nucleus. I was in the
early stages of this study when I learned during a phone conversation with Mel Newman that he had conceived of and assigned
this same problem to one of his predoctoral students, Robert Hart,
who was further along than I; hence I terminated my efforts. The
project was very successful and yielded an extraordinarily facile
entry to substances of interest in the study of carcinogens.[4] The
authors gave me a kind acknowledgment in footnote 4 as follows:
"The authors wish to express their gratitude to Dr. W. S. Johnson,
University of Wisconsin, who had independently started work
along similar lines, for deferring to our interests in the problem."

In due course, the Newmans and Johnsons became good
friends, visiting each other at work and at home, for example, to
witness the football team of Ohio State destroy that of Wisconsin.
In the summer of 1956, I spent several delightful weeks lecturing
at Ohio State, which afforded me the opportunity to see a lot of
the Newmans as well as of a favorite relative, my uncle William
Summer, after whom I was named. Mel and I were not only interested in sporty cars but loved to listen to jazz. He had many musician friends and when the occasion arose to introduce one of them
to Bob Woodward who was visiting Ohio State, Mel said, "Bob is
to organic chemistry what Louis Armstrong is to jazz." The friend
replied, "Boy, you must be one hell of a chemist."

In the early 1940s the state of the art of synthesis was
primitive. The tools of synthesis were greatly limited; for example, many of today's common practices such as metal hydride
chemistry had not yet been developed. Moreover, the methodol-

Melvin S. Newman, ca. 1965.

ogy for forming new C–C bonds was barbaric by today's standards. (In this context it is instructional to examine the early volumes of Theilheimer's treatise on synthetic methods.) More importantly, however, art was essentially nonexistent for dealing with the stereochemical problems attending the synthesis of molecules with several asymmetric centers. Neither Werner Bachmann at Michigan nor Sir Robert Robinson at the Dyson Perrins Laboratory, Oxford, the two foremost practitioners of steroid synthesis at the time, ever reached the point of designing schemes that addressed the stereochemical problem.

William Summer, Columbus, Ohio, in the 1950s.

Thus Robinson's 1953 synthesis of *epi*-androsterone (*see* subsequent discussion), which has 7 asymmetric centers (64 possible diastereoisomers), was a *tour de force* involving production and separation of diastereomers at various stages. Nothing was known about conformational principles then, and fortunately Robinson and John Cornforth had these forces working to their benefit occasionally, otherwise the problem would have been hopeless. I can remember Robinson saying that he felt there was an intrinsic tendency for reactions to give products with natural configuration, but he was right to only a small degree. Thus, certain highly unstable isomers were not produced; on the other hand, generation of a 6/5 ring fusion is likely to give the more stable cis (unnatural) material. In the early 1940s, more was probably known about the occurrence, constitution, and physiological importance of steroids than of any other class of natural products except for carbohydrates.

Rolf Huisgen, who went to immense trouble in order to assist me in procuring this Mercedes 190 SL sports car, 1958.

Sir Robert and Lady Robinson.

Emil Fischer, possibly the greatest of all organic chemists, was way ahead of the times in understanding and coping with multichirality; however, the experience with polyhydroxy compounds is highly specialized and not generally applicable to the numerous types of hydrophobic natural products like the steroids. It is remarkable that with essentially none of today's spectroscopic facilities the constitutions of so many of the steroids were well established largely by degradative and interrelationship experiments. Thus, in the 1940s a number of "partial" synthesis studies had been realized, such as conversion of cholesterol into progesterone as well as testosterone.

Total synthesis of steroids was a particularly exciting challenge for the following reasons:

1.   There were important structural challenges, particularly the difficult problem of producing fused ring systems with the angular methyl group. Methodology of general applicability to this end was nonexistent.

2.   The immense problem of introducing the many new chiral centers diastereoselectively was highly challenging and promised to generate much new methodology. Note that the Robinson–Bachmann work had to come first for this issue to be appreciated.

3.   Finally, it was exciting to have as targets those substances that have so many biologically and potentially medicinally important roles.

At that time it was known that the steroids included the male and female sex hormones, the pregnancy hormone, the bile acids, provitamin D, and the cardiac glycosides. Also, the identification and exciting biological properties of the adrenal corticoids were being disclosed from the laboratories of Wintersteiner, Kendall, and Reichstein. In addition, Russell Marker's elucidation of the structure and reactions of the saponins was just being disclosed. The next decade was to see steroids and the ubiquitous plant and animal triterpenoids linked biogenetically, and the triterpenoids became interesting, related synthetic targets. The 1940s were the beginning of a worldwide expansion of major steroid research efforts in industrial (mostly pharmaceutical) and academic laboratories. Over a half of a century later the scientific world finds this field still thriving and just as full of excitement. In the pages that follow, an attempt is made to show how a segment of the field has developed and to speculate as to where it may be going.

The culmination of our work in this early period was the stereoselective synthesis of equilenin methyl ether (**14**) and equilenin (**7**) (Scheme 4).[5,6] My extraordinarily gifted predoctoral student, Jack Petersen (Ph.D. 1946; six publications), was exploring the Stobbe condensation of desmethoxy **12** and discovered that the process in this case was more complicated, involving a Thorpe–Dieckmann reaction with the cyano group, so that the final product, much to our delight, was desmethoxy **13** having all four rings. I immediately suspended my other bench work to carry out pilot experiments in the methoxy series, and Dave Gutsche gave the Bachmann touch to the whole synthesis to obtain the results shown in Scheme 4. We were lucky that the hy-

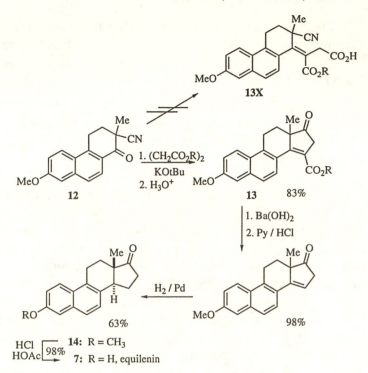

Scheme 4. The Wisconsin total synthesis of equilenin methyl ether (14) and equilenin (7).

drogenation of the enone from the decarboxylation of **13** gave predominantly equilenin methyl ether (**14**) with the natural stereochemistry. Acid hydrolysis of **14** led to the formation of equilenin (**7**). Serendipity played a strong role in this achievement because we were rather expecting ketone **12** to undergo a simple Stobbe condensation to form **13X**, a reaction that was not observed.

The Wisconsin Alumni Research Foundation, a philanthropic organization of some means (established originally on the income from the Steenbock patents on the irradiation of foodstuff) took interest in this equilenin synthesis, helped us to obtain patents, and financed a pilot-plant operation under the direction of my students, Gene Woroch and Hans Wynberg. Note that the natural product is a rarity. The annulation reaction **12** → **13** was performed on a one-mole scale, and over a kilo of racemic equilenin methyl ether (**14**) was finally produced by this synthesis. In addition Gene effected the resolution of over 100 grams of the hormone. Generous samples of these materials were given to anyone who was interested in performing biological tests. C. W. Lipman even tried equilenin for treating migraine in women. The Foundation's aim of finding something of commercial value, however, was never realized. I believe that many of these materials are still in existence and are in the custody of Al Wilds at Wisconsin.

Eventually we published a total of 25 papers on the Stobbe condensation, most of which are mentioned in the review.[3] This reaction was eventually applied to a total synthesis of estrone.[7]

Angular Methylation

My early interests concerned methods of introducing the angular methyl group into fused ring systems to give an array common to steroids and many cyclic terpenoids. I personally performed the series R = C_6H_5 shown in Scheme 5[8,9] as a model for producing the C/D ring system of 17-keto steroids.

Unfortunately, the step **16** → **17** was not stereoselective because the unwanted cis isomer was slightly preponderant as determined by column chromatography, which readily effects clean separation. Much later the cis selectivity was used to advantage in a highly stereoselective synthesis of aldosterone (*see* subsequent discussion). The methodology of Scheme 5, however, was used in the first or very early total syntheses of many steroids. The earliest example is the very short although completely non-stereoselective route to estrone shown in Scheme 6.[10,11]

Sir Robert Robinson and his collaborators at Oxford had developed a facile synthesis of 18-norequilenin (substance **7** without the angular methyl group);[12] hence, there was a need for methodology involving introduction of the angular methyl group as in Scheme 5 (**15** → **16** → **17**), followed by removal of the blocking group. Although I had worked out a method for converting **17** (R = C_6H_5) into 9-methyl-1-decalone[8], the removal of the benzyli-

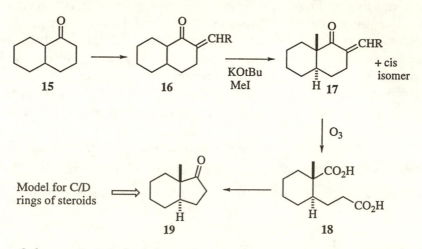

Scheme 5. Methodology for introduction of the angular methyl group into a fused-ring system. [This chemistry was "personally performed" by the author.]

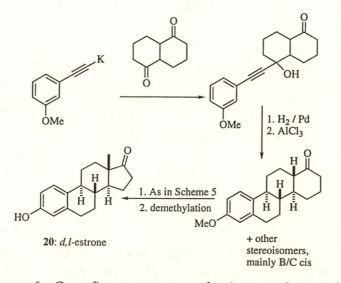

Scheme 6. Our first, nonstereoselective synthesis of estrone.

This snapshot of me in my private laboratory at the University of Wisconsin was taken in 1942 at the time I was working at the bench on the seminal studies (Scheme 5) of the angular methylation sequence. (See pages 37 and 38.)

Dilip K. Banerjee, my first postdoc (1947–1949), who was later to enjoy an illustrious career at the Indian Institute of Science in Bangalore. We were in awe of Dilip, whose biceps measured the same as Jack Dempsey's. Dilip's response to our reaction was to tell us about his uncle who was "really strong" and had a worldwide reputation fighting wild tigers with his bare arms. Dilip participated in the synthesis of estrone (Scheme 6).

dene group was awkward. Therefore, I suggested to a new student, Harvey Posvic (Ph.D. 1946), that he explore the sequence in which R = alkoxy with the expectation that the alkoxymethylidene blocking group would be readily removed by hydrolysis. Eventually he nicely reduced this plan to practice.[13,14] Before our publication, Arthur Birch, working in Robinson's laboratory at the Dyson Perrins Laboratory in Oxford, had developed the use of the *N*-methylanilinomethylidene blocking group, which similarly could be removed hydrolytically.[15] Hence Arthur and I began corresponding, and we have since had a very friendly association, having met on numerous professional occasions in various parts of the world. Arthur has published his autobiography, *To See the Obvious*, within this Profiles, Pathways, and Dreams series.

My first personal contact with Sir Robert Robinson developed in a rather unusual way. At the beginning of his project, Harvey Posvic located a paper of King and Robinson[16] describing the rather surprising reaction of cyclopentanone with ethyl orthoformate in the presence of sodium ethoxide to give 2-ethoxymethylidenecyclopentanone. Harvey tried the reaction with decalone **15** but none of the desired product **16** (R = OEt) was produced. Therefore, he repeated the published work with cyclopentanone, and he isolated a product as well as its semicarbazone with the described properties. However, this material proved to be cyclopentylidenecyclopentanone. The Robinson paper also described the reaction of the presumed keto enol ether with methylmagnesium iodide as giving "a *hydrocarbon*, C_8H_{12} ... and not the expected carbinol." This hydrocarbon, of course, was surely $C_{11}H_{16}$, with which their reported CH analysis was in fair agreement. With Harvey's approval, I wrote a letter to Robinson describing Posvic's findings indicating that we did not intend to publish them and suggesting that he use the information to set the record straight in any way that he wished. In a footnote on page 584 of his paper XLIV on steroid synthesis,[17] Robinson wrote in his inimitable style:

> Professor W. S. Johnson and his collaborator, H. M. Posvic, have kindly informed us that, under the conditions described in Part XXXV (J., 1941, 467), no condensation occurred between *cyclo*pentanone and ethyl orthoformate and that the only product was *cyclo*pentylidene*cyclo*pentanone. This finding is in agreement with the only analysis quoted (%N in a semicarbazone) and with the properties of the supposed ethoxymethylene*cyclo*pentanone (cf. Kon and Nutland; J., 1926, 3011). In view of the exiguous evidence and the inapplicable catalyst we had

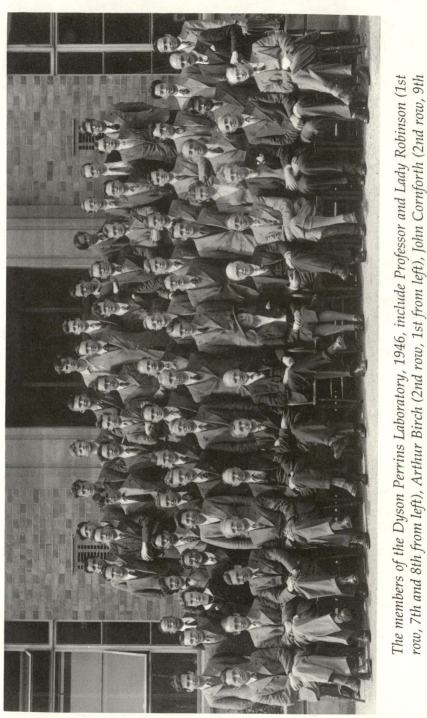

The members of the Dyson Perrins Laboratory, 1946, include Professor and Lady Robinson (1st row, 7th and 8th from left), Arthur Birch (2nd row, 1st from left), John Cornforth (2nd row, 9th from left), and Rita Harradance (later to be Lady Cornforth; second row, 12th from left).

> intended to expunge the record of this experiment
> but the matter was unfortunately forgotten.—L. E.
> K., R. R.

After that Sir Robert was always quite friendly to me. We saw each other occasionally, mostly at meetings, and twice in our respective homes. Lady Robinson, also a chemist who still did benchwork at Oxford, was a warmer person than Robert and the only time I recall seeing him glow was when I played Respighi's *La Boutique Fantasque* on my hi-fi system for him. At the same time, however, he appeared to be a bit annoyed because my equipment sounded better than his.

I quote from parts of his letter of January 7th, 1959: "First to say that our time in Madison was one never-to-be-forgotten and a most pleasant experience in every way. Particularly we are deeply grateful to you for the great hospitality you showered on us and the insight you gave us into the research in progress & the working of the Department. ... I have one sad memory and that was my defective gramophone reproduction. This will cost me dear when I am quite sure how I can recover this bit of self-esteem. But at the moment I doubt whether equipment equal to yours is available here."

The sound equipment that I was using when Robinson was so impressed was state-of-the-art at the time (before stereo) and it included the legendary Klipschorn. This very expensive speaker involves a folded exponential horn that was designed and patented by Paul Klipsch. When it is placed in the corner of a room the walls act as an extension of the horn yielding a realistic low and mid-range bass response that has no peer even to this day.

In the 1950s a "K-horn", which was and still is manufactured in Hope, Arkansas, was hard to come by. On discovering that the nearest dealer was in Chicago, I wrote to Klipsch suggesting that he consider making me his temporary representative in the Wisconsin area. The upshot of this venture was that before Paul and I had met each other, we worked out an arrangement whereby he provided me with a Klipschorn at the dealer's discount and I made myself available to demonstrate the unit in my home to anyone who was interested. It was understood that I would not be expected to do any active promotional selling. Thus I obtained my first unit early in 1954. As it turned out, I had a lot of fun making a few sales exclusively to my colleagues at the University of Wisconsin, to whom I passed on my discount. Also eventually I had the pleasure of meeting Paul Klipsch, a very colorful character, on several occasions. He was particularly interested in my move to Stanford where he had obtained a master's degree in electrical engineering under Fred Terman, who was later to become my boss at Stanford.

It was this demonstration equipment that had made Sir Robert so unhappy. A few years later, in 1956, Robinson wrote a letter to one of his students, John Pike, who was doing postdoctoral work with me. The last paragraph of this letter, which John kindly transmitted to me, reads as follows: "With my kindest regards to Professor Johnson and by the way, please tell him that the injury he did me by proving that he had a better gramophone than I, has probably been put right. But I expect he has changed his apparatus too. Further, I trust he has now demagnetized the tape recording containing some impressions of Sir William Pope." I never did have the heart to advise Sir Robert that I had indeed made a significant improvement in system by replacing the medium-to-high frequency part of the K-horn with a JBL 375 theater horn unit with acoustic lens.

It is my feeling, from observing chemists, that people who have exalted opinions of themselves are generally unhappy individuals. Sir Robert was a rather extreme case. He was compulsively competitive, and had almost paranoid delusions that others were trying to steal his ideas. Without finding out anything about Bob Woodward's steroid synthesis, Robinson accused him of thievery. Bob, who related this to me, seemed to be more amused than hurt. Of course, Robinson was to learn that the Woodward synthesis had no similarity to his own and he ended up bearing no grudge. Robinson's troubles with and vendetta against Sir Christopher Ingold are well known;[18] for example, Sir Robert devoted the major part of a lecture he gave at Stanford in the 1960s to making a case against Ingold for appropriating his ideas on the electronic theory of organic reactions. What a pity that this man could not relax and enjoy his own great talents. For more discussions on the Robinson–Ingold relationship on the development of

conformational analysis and steroid and natural products synthesis and chemistry, the reader is encouraged to read Derek
Barton's autobiography *Some Recollections of Gap Jumping* within
this Profiles, Pathways, and Dreams series.

Some Incidental Studies

Intramolecular Acylation

During my Ph.D. studies I developed a special interest in intramolecular Friedel–Crafts type of acylation reactions, and in the
early Wisconsin days, a few projects were explored. Also at the
request of Roger Adams, I wrote a chapter on the subject for the
second volume of his (then) new series entitled *Organic Reactions.*[19] An example of particular interest was the cyclization of the
ester **21** (Scheme 7), which is readily prepared from reaction of *m*-
chloroaniline and *n*-propyl acrylate. The product **22**, on treatment
with the diamine **23** in nitrobenzene gave chloroquine (**24**), which
at that time was a very important antimalarial drug manufactured
by the Winthrop Chemical Co., a division of Sterling Drug. This
synthesis was based on chemistry developed by Gene Woroch,
Benny Buell, and Bill DeAcetis as part of their Ph.D. disserta-

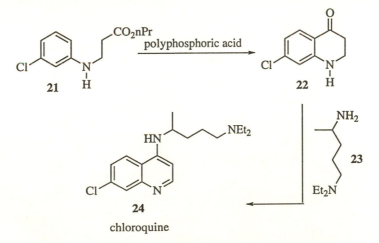

Scheme 7. Synthesis of the antimalarial, chloroquine.

tions.[20-22] The Sterling-Winthrop Research Institute, for whom I was a consultant, submitted our synthesis to evaluation by their product development division. As it turned out, to realize high yields at the cyclization stage, impractically high dilution was required; otherwise, the very short synthesis probably would have replaced the existing commercial method.

Neighboring Group Effects

A small but at the time significant study came about in a rather unusual way. After the death of Homer Adkins in 1949, his students were divided among the organic faculty. One of three assigned to me was Elliot Schubert, a fourth-year graduate student who was rather desperate for lack of positive results in his hydrogenation project. There was some question about the identity of the 2-aminocyclohexanols resulting from hydrogenation of 2-oximinocyclohexanol, and I suggested that he concentrate on the stereochemical aspect of this problem. We both soon became extremely intrigued, for a vein of gold appeared in examining the reaction of thionyl chloride with the *cis*- and *trans*-2-benzoyl-aminocyclohexanols. Schubert showed with the *trans*-isomer (Scheme 8) a stereospecific inversion was involved and, for the first time, an intermediate was isolated (in this case the oxazoline)

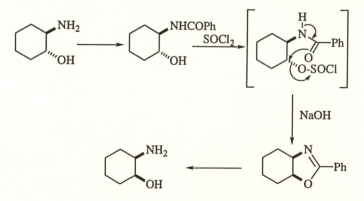

Scheme 8. Isolation of an oxazoline as the intermediate in the neighboring group participation of the benzamido group in the substitution of a vicinal hydroxyl group.

of the type postulated by Saul Winstein and others for neighboring group effects in displacements. This chemistry is directly related to the interconversion of threonine and *allo*threonine. Schubert completed this study in about six months, received his Ph.D., and we wrote a full paper on the subject. Winstein, who refereed our paper,[23] had just completed a similar study and asked if we would wait for him to prepare a communication to the editor[24] and of course we agreed. In the meantime I was asked to review a manuscript by McCasland and Smith[25] involving a very similar study; so it was arranged that all three papers would appear in the same issue of the *Journal of the American Chemical Society*.[23–25] Thus Schubert and I found ourselves involved in a very active area.

At this point I would like to digress briefly to comment on the attitude many people in my profession seem to have regarding the influence they purport to have had on predoctoral students. It is nonsense for professors to regard themselves as Pygmalions who create exceptional scholars. I submit that those of us who have had occasional students who become distinguished scholars are simply lucky and that such students will not be denied; they will turn out just as well no matter who they work for. There are some special situations where the professor's influence

Another person who had considerable interest in this work and who helped us considerably in the interpretation of the rate studies was the late Saul Winstein pictured here. Aside from being one of the truly great physical-organic chemists of all time, he was a fine person. Indeed I almost felt badly when I won a case of Jack Daniels as the result of a bet he and I had as to whether Henry Taube would move (from the University of Chicago) to UCLA or Stanford. (Photo courtesy of Sylvia Winstein.)

can be important, and the case of Elliot Schubert is an example. I entered his career at a propitious point, and gave him some encouragement. He had already developed considerable experimental expertise so that all he needed was to be pointed in the "right" direction. In such situations the student is likely to exaggerate the importance of the professor's part in the drama. Thus Elliot has ignored my feet of clay to this day. I hear from him regularly, and he repeatedly speaks of the influence I have had on his life. It is rather frightening to find oneself involuntarily having to play the role of a divine power.

The 1950s

This was a very exciting period that brought us the concepts of conformational principles and NMR spectroscopy.

Conformational Studies

Derek Barton visited us in Madison in 1950, just before the appearance of his classic paper on conformational principles,[26] and one evening he gave a private seminar to Al Wilds and me on the subject. That occasion suddenly lifted a heavy veil of mystery that surrounded much of the chemistry in which we were involved, and I immediately became a Barton disciple and spread the gospel at home and around the country. This was the beginning of a warm friendship between the Bartons and Johnsons, which has lasted to this day. Considering how far apart our homes have been, we have gotten together socially as well as professionally on numerous occasions. Derek is especially fond of cats and has always inquired about ours. In May 1975, when he and Christiane were staying with us, I showed them how our cat would stand on his hind legs waiting for a treat. When I asked Derek what he thought of the cat's performance, he said, "Near genius."

I was so excited by conformational analysis (this term was coined by Barton) that we were inspired to undertake a number of studies in the area, publishing 10 papers altogether, the first in 1951.[27] The most significant of these contributions was the first experimental determination of the difference in enthalpy (5.5 kcal/mole) between the chair and boat forms of cyclohexane.[28] This was accomplished by combustion calorimetry (performed by my colleague John Margrave) on a pair of lactones (*see* Scheme 9)

Derek Barton with a friend, ca. 1961. (Photo courtesy D.H.R. Barton.)

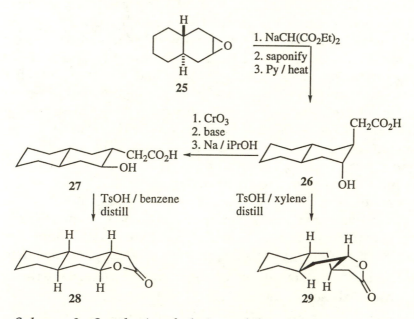

Scheme 9. Synthesis of chair and boat isomers used for combustion calorimetry.

that differed only in that the central ring of one (**28**) was in the chair conformation and the other (**29**) was in the boat form. The synthetic work was carried out by Victor Bauer who produced the boat isomer **29** by lactonization of the diaxial hydroxy acid **26** under forcing conditions. The diequatorial hydroxy acid **27**, as expected, was easily lactonized to give **28**. It was in the full paper[29] that the term "twist" boat was coined.

The Hydrochrysene Approach to Steroids

This developed into a long and fruitful project lasting for more than 10 years. The strategy involved facile production of a tetracyclic product **34** having the C-19 angular methyl group (Scheme 10), then stereoselective reductive generation of asymmetric centers in the nucleus, followed by introduction of the C-18 angular substituent (*see* previous discussion). Compared with other meth-

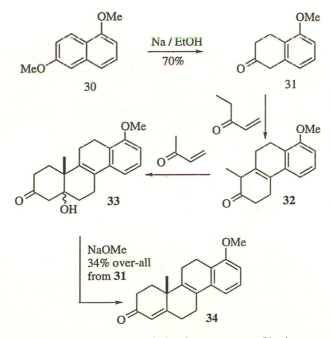

Scheme 10. Preparation of the key tetracyclic intermediate in the hydrochrysene approach to steroids.

At a lecture in the early 1950s, discussing conformational analysis, hindrance in the cholestane and coprostane series, and ease of diaxial elimination.

ods of that time, this "new" approach to steroids proved to be relatively short and highly stereoselective, although it is not so impressive when compared with the current state of the art.

The steps to products **31** and **32** were based on chemistry of Cornforth and Robinson.[30] The key transformation **32** → **33** was an application of what I call the "Wilds' principle of vinylogous activation" for adding a new ring.[31] Al's example of this ingenious method of controlling the regiochemistry of the Robinson annulation is shown in Scheme 11.[32]

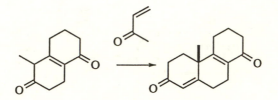

Scheme 11. Wilds' principle of vinylogous activation of the Robinson annulation.

The tetracyclic substance **34**, with one asymmetric center, was easily made in kilogram quantities. Prior art indicated that there was a good possibility that alkali metal in ammonia reductions of the olefinic bonds in the molecule should favor the thermodynamically more stable diastereomers. After extensive study we learned how to introduce new chiral centers by stereoselective reduction of the olefinic as well as the aromatic bonds. Vicinally trisubstituted aromatic rings are notoriously resistant to Birch reduction, and after numerous attempts we had given up on applying this method to reduction of the aromatic ring D, because our best yield was about 10%.

When Brian Bannister, one of Sir Robert Robinson's last Ph.D. students, arrived for postdoctoral work, he had heard of some unpublished experiments of W. F. Short on forcing conditions for the Birch reduction involving the use of high concentrations of lithium and ethanol. Despite my skepticism he wanted to try out this idea on our system and, to my delight, proceeded to make the reaction work quite well. Submission of **34** (*see* Schemes 10 and 12) directly to the Bannister–Birch reduction conditions gave **36** along with the regioisomeric α,β-unsaturated enone, and H$_2$/Pd reduction of this mixture of enones gave **37**. Thus the two tandem reductions yielded, stereoselectively, a tetracycle with 6 new asymmetric centers. This was a major breakthrough and soon

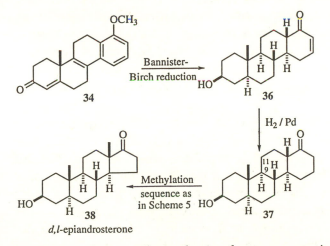

Scheme 12. Our first total synthesis of a nonaromatic steroid.

led to our first synthesis of a nonaromatic steroid, *d,l*-epiandro-sterone (**38**) (Scheme 12) with two angular methyl groups.[33,34] This work was completed later than that of Cornforth–Robinson[35] and of Woodward,[36] but despite the lack of stereoselectivity of the angular (C-18) methylation step, ours was more stereoselective overall and relatively short.

On a personal note, Brian was a dedicated night owl. His predilection for working at night became almost an obsession. He claimed that he worked much more efficiently this way because equipment such as the NMR and other spectrometers were more accessible. Thus I often would get a progress report from him when I arrived at the lab in the morning, just before he left the lab to go home to sleep. In this fashion, he was sensationally productive; otherwise, I could not have tolerated such behavior. When it came to a permanent job, I warned him that he would have to change his ways. The only place that looked hopeful was Upjohn where the recruiter said they were not critical about working hours, so he accepted their offer. As it turned out, when he started work and brought the matter up again, he was told that he could arrive or leave five or ten minutes early or later without any problem. Be that as it may, Brian proved to be one of Upjohn's most productive employees.

Our synthesis of a steroid was a completely total synthesis, whereas the earlier ones were "formal" total syntheses. The formal term was coined by Cornforth and Robinson for a synthesis that depends on "relays" through intermediates that were supplied (for the further steps) by degradation of natural steroids. Thus their first relay was a tricyclic material having rings A, B, and C.

Woodward would never use the term formal total synthesis; indeed, he steadfastly refused to recognize that there was any difference between a synthesis that was carried all the way through from simple chemicals and one that relied on relays. Bob and I used to have rather heated arguments on this issue, presenting each other with absurd hypothetical cases, such as if you effect a completely total synthesis of testosterone acetate, then saponify the acetate prepared from natural material, are you obliged to label this as a formal total synthesis of testosterone?

Brian Bannister (postdoctoral associate 1954–1955), who made our first non-aromatic steroid synthesis possible.

Neither of us ever convinced the other, and I still feel that when a significant number of the steps of a synthesis are performed using material derived from the natural product (i.e., by a process commonly referred to as "partial synthesis"), then it is not right to call your final product "totally synthetic material".

It is hard to describe the high degree of excitement that pervaded our research group during the late stages of the synthesis of the steroid **38**. The trans-anti-trans-anti-trans stereochemistry of this product was based only on the presumed stereochemical course of each step of the synthesis, and therefore could not be assured until a comparison was made with the natural product. Barbara and I customarily had a Christmas Eve party at our home for those in the group who could not get home. On that occasion, in 1951, the group presented me with a wire (the forerunner of tape) recording consisting of a skit about the Johnson lab. Barry Bloom (postdoc. 1951–1952, now President of Pfizer Central Research) portrayed me in the skit, and I was told on good authority that he also prepared the script and directed the production. It began with a cheer:

> Trans-anti-trans-anti-trans
> We've got about a 50-50 chance
> If we don't get trans-anti-trans
> We're liable to lose our pants, Hey!

I was given a good roasting with such incidents as:

> *Johnson (Bloom) entering one of the labs*: Hi, Ed, how are things going?
>
> *Ed (Hans Wynberg)*: Well, Sir, since you were here 10 minutes ago, I have filtered the solution and ... etc.

It was rumored that because of my reputation for frequent lab visits and expecting my collaborators to work hard, these people referred to me as "Black Bill". During one of my consulting visits at Sterling in the 1950s, Syd Archer arranged for a celebration during which he presented me with an authentic 10-foot leather bull whip. The whip, which I have always kept on display in my office, has perpetrated more than just comments. On one occasion, Jim Korst, who was scheduled to deliver the departmental organic seminar, gave his performance while wearing a torn white shirt covered with "blood" stains.

This was an exciting, happy, and playful period, and I feel that Barry's sense of humor was contagious. At that time he had set up a huge Hanovia quartz lamp to irradiate epiandrosterone to give the C-13 epimer for comparison with synthetic material also produced in the sequence **37** → **38**. The apparatus was en-

1951 Fasnacht Celebration, Basle, Switzerland. One of the masqueraders comes on to an observer at the concealed displeasure of the latter. When the editor of this series originally suggested, "Why is this man smiling?" as a caption, Bill responded, "I assume this is a rhetorical question."

shrouded by long black curtains, on which signs began to appear. One read: "Genes Mutated While You Wait."

I now digress briefly to report a shocking incident that occurred in our laboratory. In September 1951, John Smith (a fictitious name) began his fourth year of graduate study with me. He was an A student, had passed all of his comprehensive exams, his research was going very well, and he was nearing the point of writing his Ph.D. thesis. John was a lone wolf, very quiet, abrupt, and occasionally sullen in attitude. I always felt a bit uncomfortable when talking to him.

One day in the early winter Allan McCloskey (Ph.D. 1951), Dick Shafer (Ph.D. 1951), and Hans Wynberg (Ph.D. 1952) came to my office and announced that they were greatly concerned about John because he seemed very withdrawn and unhappy; in addition, they had either witnessed or were the objects of serious threats of violence made by John. The four of us met with a school psychiatrist who said that John sounded paranoid, but that nothing could be done unless he volunteered for help or perpetrated an act of violence. I was urged to try to induce John to talk with the psychiatrist, but all I could do without possibly further jeopardizing those who were threatened was to say to John that he seemed to me to be unhappy and perhaps he could benefit from some professional help. He scoffed at the suggestion. The following account of the blowup is based on information I eventually collected from all involved as well as authorities who dealt with John afterward.

On February 29, 1952, John came to the laboratory in a state of agitation and also, apparently, inebriation. That afternoon he told Bill Loeb (Ph.D. 1954), one of his labmates who John always rather liked, that he (Bill) ought to go home. Unfortunately for Bill, he ignored this warning. It was later learned from John that he had planned to shoot Hans Wynberg with a .22-caliber revolver he had bought for this purpose, after Hans' Ph.D. oral, which had been held on the previous day. It seems that Wynberg was number one on the death list because he had tried on various occasions to make friends with John and to bring him out of his shell. John interpreted these overtures as interference and thought that Wynberg had ulterior designs. Besides, Hans, who had begun his Ph.D. work at the same time as John, was an object of jealousy because he was finishing first. As it turned out, Hans never returned to the lab after his exam, as he had planned to leave town immediately. By late afternoon, when John realized that his plans had been thwarted, he went berserk. Bill Loeb's account of the incident, written on June 7, 1987, follows.

> I heard a bang. I turned around and saw [John] with a kind of dreamy look, point his revolver at a dish high on a shelf and fire again. Like others I have read about, I thought that he must have been shooting blanks. And I asked him, 'Are those blanks, John!' He then fired at me. I felt something hit my right chest, but I still did not believe that I was threatened. I put my left hand to my chest and looked down. At this point John fired again, hitting me in the left hand and I saw blood. The light finally dawned. ... I ran out of the lab, meeting Barry Bloom, Hank Dehm, and Bill DeAcetis coming out of the lab next door. 'John shot me!' I believe I said. I headed for the north end of the laboratory building, went down those three flights of stairs, pulling along Barry and Vic Soukup, whom we had picked up along the way, across the street, through the hospital, into the infirmary, up one flight of stairs and onto a bed on the second floor. It turned out that the first bullet passed through the right lung and was found in my shirt. The second, which went through my left thumb and along a rib is still in my right chest.

Many of the people in the lab panicked for by that time they had been alerted to the threats and possible violence. Barry Bloom, for one, kept his cool and helped Bill Loeb over to the hospital (Barry's version was that despite spurting blood Bill was running so fast that it was hard to keep up with him). Others dispersed in various directions. Mac McElvain, whose office was on the other (east) side of the building saw someone running in the hall and asked what had happened. The runner replied that there had been an accident in one of my labs. Mac, thinking it was chemical in nature, rushed over to find John alone in his lab trying to reload his revolver. When he asked what had happened, John pointed the gun at Mac who quickly left. Luckily, John had been unable to remove the shell cases after the revolver was fired.

In the meantime the police and the press had arrived. They wisely refused to go into John's lab and eventually forced him out with tear gas. John, assuming that I had summoned the police, was shouting, "I'll get Johnson," which was highlighted in press releases all over the country. The fact is, I knew nothing of the incident until John was incarcerated, because I was at home preparing to leave on a consulting trip. I received accounts on the phone and met with the district attorney in the railroad station before my departure. My main concern was for Bill Loeb, and I would not have left town except that I was assured that he was not in danger.

The evening of the shooting John tried to kill himself by running into the walls of his cell. He was then transferred to the mental hospital where he was adjudged "a paranoid type *dementia praecox*", whereupon he was committed to the state prison for the insane where he remained for about a year. He and I communicated some during that year, and I arranged for him to receive an M.S. degree. He was eventually declared "cured" and released in the custody of his sister in Maryland. He has not been heard from since.

After the incident, Bill wrote me a letter that included the following anecdote:

> Marty Swarts, a friend and a plant physiology grad student living at the University Club, had been away from Madison all day. He returned to his room at about 11 p.m. His roommate said, 'Did you hear the news? Bill Loeb shot somebody.' Marty was stunned. As he told me later, he sat up for a long time trying to scheme out what could have caused such an apparently friendly and stable person to do this. He said to his roommate, 'Bill got married two months ago. Do you suppose that was it?'

Anyhow, work went on and the hydrochrysene methodology was applied to the synthesis of corticoid intermediates, and the 11-hydroxy group was introduced by the steps shown in Scheme 13.[37,38] This approach probably would never have been realized had it not been for Gilbert Stork's help and encouragement. He gave us full particulars (unpublished) relating to similar chemistry that he had developed in his laboratory. This is only one example of the great inspiration I have derived from contact with Gilbert during more than 40 years. I am very lucky to have as a close friend a person who not only is one of the really great scientists of all time, but is an articulate philosopher, a lover of art, and an incomparable humorist. Gilbert, Winifred, and the two of us have had some memorable times together in various parts of the world. Incidentally, I feel that I would almost certainly not

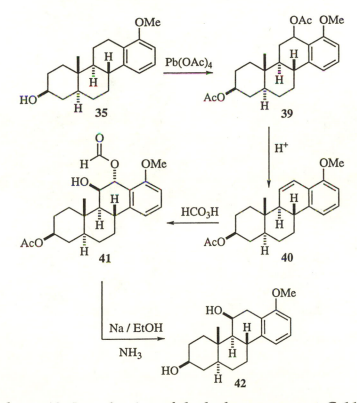

Scheme 13. Introduction of the hydroxy group at C-11 en route to corticoids.

have been the recipient of the first Roussel Prize (in 1970), had it
not been for Gilbert's cogent effort to bring this about.

Gilbert is the funniest person I have ever known. There is
nothing contrived about Gilbert's humor, which just comes naturally, and being with him engenders a feeling that is not unlike
watching a Woody Allen movie. Several chemists collect and ex-
change anecdotes about him; one of these is recorded here. On the
occasion of the 1957 Spring ACS meeting in Miami, Gilbert was
receiving one of the most prestigious honors in chemistry, the
ACS Award in Pure Chemistry. The Storks and Johnsons had arranged to stay at a hotel at Miami Beach. It was very hot and we
got badly sunburned before the meeting. Gilbert had rented a

*Gilbert and the late Winifred Stork and Edith Roberts, wife of
Jack Roberts. According to Gilbert Stork, "We're all dressed up
and ready to go, around 1951. Jack [Roberts] thinks we are on
the way to the [John C.] Sheehans' for dinner. Obviously some
sort of celebration, as Edith has an orchid." (Photo courtesy J.
D. Roberts.)*

convertible for taxiing over to the city where the sessions were being held. While driving over, with the top down, just before his award address, he kept looking at some rather crumpled papers which he propped up on the steering wheel. When questioned, he put on an air of nonchalance in the face of utter disaster and explained to us that he was trying to decide what he was going to talk about. The very large auditorium was packed with people, most of whom had, not long before, heard a talk given by Bob Woodward who appeared, as usual, immaculately dressed in his blue suit and began his talk with the dramatic introduction: "The lecture that I am privileged to deliver today concerns recent work that has never before been disclosed in the Western Hemisphere." Now Gilbert, after being introduced, stood up at the podium looking quite non-Woodwardian in his rumpled suit that had suffered from the open air ride in the severe heat. Then he began "The lecture that I am privileged to deliver today concerns recent work that has never before been disclosed in Miami." This brought the house down, and I laughed so hard as to cause conversion of an incipient hernia into a major rupture requiring surgery soon after I returned home. (Before writing the above anecdote, I phoned Gilbert to see how he felt about having it published. Among other things, he said, "I never did understand why people thought my remark was so funny.")

Continuing the hydrochrysene story, eventually we found one quite gratifying solution to the problem of unfavorable selectivity in the angular methylation step. By using a substance like **37** (Scheme 12) but with a 9,11-double bond in ring C, the methylation of the furfurylidene ketone gave predominantly the C/D trans product, isolated in 69% yield.[39,40] This result was all the more rewarding because 9,11-dehydrosteroids are useful intermediates for making corticoids. This principle for favoring the trans methylation product was also applied to a completely stereoselective total synthesis of estrone, the required 9,11-dehydro intermediate being obtained very conveniently by a Diels–Alder reaction of 1-vinyl-6-methoxy-3,4-dihydronaphthalene with benzoquinone. When my student John Cole (Ph.D. Wisconsin, 1960) was well into this study we learned that James Walker (National Institute of Medical Research, London) with his collaborator P. A. Robins was also well along with the same

Gilbert Stork and Derek Barton, two very important people in my life. (Photo courtesy of Carl Djerassi.)

study. Therefore, we decided to join forces and publish the work jointly.[41,42]

Other total syntheses accomplished by the hydrochrysene approach include testosterone,[43–45] cholesterol,[46,47] conessine,[47,48] progesterone,[47,49] aldosterone,[50,51] and later, veratramine.[52–54] Except for veratramine, which relied on a relay, all of these were completely total syntheses.

The synthesis of aldosterone deserves special mention because it was stereoselective throughout. The substance **43**, which we had used for introduction of the angular methyl group to produce an 11-oxygenated steroid, was alkylated in a Michael reaction with methacrylonitrile to give exclusively the *cis*-fused C/D

This is the house we built in the Madison Arboretum on Balden Street, about 1957. The street number plaque with the brass steroid insignia was swiped, shortly before our move to California, presumably by one of my students who made restitution some 20 years later by arranging for the plaque to be placed surreptitiously at my seat while I was at the podium during the memorial ACS symposium for Robert Woodward in New York, 1979. I take this opportunity to thank whomever was responsible for the safekeeping and return of the plaque.

ring product **44** (*see* Scheme 14). (The high cis selectivity compared with the methylation reaction was unexpected.) Now the
cyanoalkyl group was destined to become part of the D-ring of
the final product *d,l*-aldosterone (**47**), whereas the keto group of
the D-ring in **44** was to become the angular aldehydo group in **47**.

Ours was one of four total syntheses of aldosterone that
appeared within a 2-year period.[55] All of these syntheses were
long and none could compete with Barton's[56] partial[57] synthesis of
aldosterone starting from corticosterone, which is quite efficient.

I was scheduled to deliver a talk at the 1958 Spring ACS
meeting in San Francisco on the occasion of receiving the ACS
Award for Creative Research in Organic Chemistry, and I had

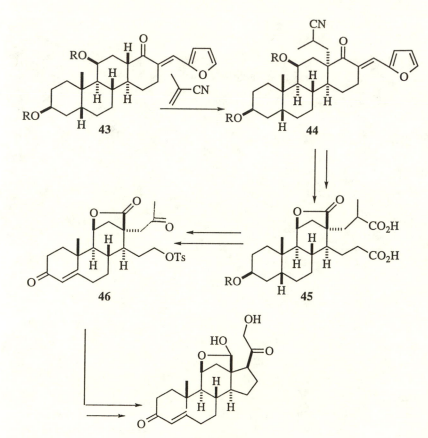

47: *d,l*-aldosterone

Scheme 14. Wisconsin synthesis of aldosterone.

decided to give a progress report on our aldosterone synthesis, which was just nearing completion. Mordecai "Mort" Rubin (postdoc. 1956–1958) had prepared the first specimen of what we presumed to be racemic aldosterone; about 1 mg was isolated by preparative paper chromatography, and we mailed this specimen off to my old Harvard friend, Norman Jones, at the Canadian Research Council, who had the equipment and expertise to obtain solution IR spectra on a micro scale. On the day before my lecture Mort phoned me in San Francisco announcing that we had received a telegram from Norman reporting that the spectrum of the synthetic sample was identical with that of natural aldosterone.

Naturally, I was delighted to receive the aforementioned award (I was the second recipient, Bob Woodward having been the first). However, I had already, in 1957, received one of the greatest honors I could imagine: an invitation to give the first Werner Bachmann Memorial Lecture at Michigan. I am sure that my former student, Bob Ireland (Ph.D. 1954), who was then on the Michigan faculty, was to a large extent responsible for my receiving this invitation. On the night following the lecture, Bob and I enjoyed an all-night celebration that included among other things a bizarre incident involving an abortive attempt on our part to recover a lost dog belonging to Leigh Anderson, the Chairman of the Chemistry Department. The snapshot shown on page 68 was sent to me by Leigh who appreciated our intentions although we awakened him at 4 a.m. to present him with a look-alike boxer we had found straying in downtown Ann Arbor. Even Leigh thought the dog was his until it refused to respond to his advances.

Since 1957, Bob Ireland and I have kept in close touch. Barbara and I particularly enjoyed traveling with Bob and his second wife, Susan, into Central and South America in 1968. For some reasons that I shall never fathom, Bob has always seemed to regard me as a father, and I was deeply touched when he dedicated his first book[58] to me as well as to his real father.

A Year at Harvard

In 1954–1955 I took leave of absence from Wisconsin to spend a year at Harvard as Visiting Professor to teach in place of Louis

Bob Ireland, 1957, with Leigh Anderson's real dog, not the stray which Bob and I picked up and delivered to Anderson about 4:00 am the "night" after the lecture. The back of the photograph is inscribed, "Just a memento of a previous occasion when you gave a lecture. This is what you should have found. The model is nearer what you did find. Be careful after your lecture today. Congratulations!! Leigh"

Fieser, who was thus relieved of his classroom duties. I taught the beginning full-year course in organic (Chem 20) to a class of about 300 very smart, exuberant students who constantly overreacted to my stimuli, by hissing whenever I announced a quiz, either laughing uproariously or hissing at my jokes, and clapping when they enjoyed a presentation. (At Wisconsin the students were less demonstrative.) I found this experience enjoyable and challenging. Luckily I had Jay Kochi as my assistant who also took full charge of the laboratory and gave occasional lectures when I was out of town. In addition I gave a one-semester graduate course on steroid synthesis.

This was a busy year, because I was much concerned about my research program, involving 15 co-workers, at Wisconsin. Fortunately Raphael Pappo (now retired, Searle), an exceptionally talented chemist who had been with me as a postdoctoral collaborator since 1952, agreed to stay on as Lecturer at Wisconsin during 1954–1955 to give day-to-day supervision to the program. Also that year I was involved in completing 10 full papers reporting the work of 14 of my collaborators on much of the hydrochrysene work. Eventually all of these papers appeared consecutively in the *J. Am. Chem. Soc.*, **1958**, *78*, 6278–6361. It is noteworthy that Raphael was co-author of 16 papers from his 3-year stint at Wisconsin. After leaving Wisconsin, Raphael has enjoyed a brilliant career as a pharmaceutical chemist and has made many important discoveries.

The most rewarding aspect of that year was the personal contacts I made, particularly with Paul Bartlett, Konrad Bloch, Alex Todd (who was visiting MIT for a semester), Frank Westheimer, Bob Woodward, and rekindled friendship with Louis and Mary Fieser. Bob and I had known each other since my graduate school days and because of common interests it was natural that we spent a great deal of time together. Most of the time we discussed science, but we joked and laughed a lot, too. Jim White tells of an occasion when Dodie Dyer (Bob's secretary) told him that Professor Johnson was in the office with Professor Woodward, but he could go on in. Jim was expecting to find us deeply involved in scholarly discussion; however, what he observed was the two of us competing to see who could toss the largest number of playing cards into a hat on the floor. (The interested reader is urged to see the photo of W. Moffitt and R. B. Woodward on page 60 of Carl Djerassi's autobiography in this series, *Steroids Made It Possible*.)

As my year at Harvard was nearing its end, Bob suggested that it would be nice if something could be worked out so that he and I could be permanent colleagues, and I certainly agreed. However, the idea never reached the point of serious negotiation because neither of us was ready to leave our "present" positions. I think it was in anticipation of a development of this sort that Wisconsin, shortly before I left for Harvard, appointed me as the Homer Adkins Professor, which was established as the first research professorship in chemistry without any classroom teaching

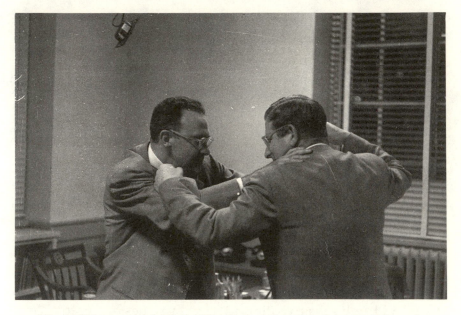

W. S. Johnson and R. B. Woodward "engaged in a vigorously exhilerant scientific discussion, Harvard, 1955. Bob suggested the idea in principle and we worked out the pose together."

requirements. I was not inclined to give up this attractive position, and Bob was strongly wedded to Harvard.

    That year I learned first hand of Konrad Bloch's classical experiments that made it possible for him and Bob to make the first compelling case for the biological conversion of squalene to lanosterol.[59] Although I was tremendously impressed by this revelation, I had no idea that it was to afford me the inspiration for a major study of polyene cyclizations that commenced in 1960 and was to last for more than 30 years.

    Thus my year at Harvard was not only very pleasant but highly rewarding. I was, however, eager to return to my home and my research group. On the other hand I knew that I would really miss the close association with Bob, who had expressed reciprocal feelings. In the years that followed we did keep in touch mostly by phone, which Bob called "the Ameche". We also had some amusing correspondence frequently involving Alex Todd. In an article Todd wrote about me in 1971, this interaction is described as follows.

From left to right: Carl Djerassi, Michail Shemyakin, Derek Barton, Robert B. Woodward, Alexander Todd, Vladimir Prelog, and Joshua Lederberg. This photo, which has appeared in previous volumes of the Profiles series, is reproduced here again for a number of reasons. First, according to Djerassi, this photograph was taken "at Stanford in 1961 (right in front of our Chemistry building) at the symposium I organized with Bill Johnson to celebrate the opening of our building." Second, Djerassi, Barton, Woodward, Todd, and Lederberg each had a major role on Bill Johnson's career. And finally, Lederberg had been misidentified as Melvin Calvin in the previous appearances of this photograph, and the site was also incorrectly attributed to Zurich during an IUPAC meeting in 1955.

It is sometimes diffiicult, if not impossible, to identify with absolute certainty the individuals in a photograph which, as in this case, was taken nearly 49 years ago. While recently autographing a copy of this photograph for Jeff Seeman, Djerassi asserted with vigor, as is his style, that the figure on the extreme right is Lederberg and not Calvin. Sadly, Calvin had earlier passed away, though not before autographing a few copies of this photograph for Seeman.

Lederberg agrees that the picture "could well be me; and that does look like the Chemistry building at Stanford, but I won't guarantee it. ." Barton identifies his "old school tie" and further asserts that "By 1961, I was not wearing it! " Seeman hopes soon to have Lederberg autograph this photograph, previously autographed by Calvin, thereby covering "all bases." Note that Barton, Woodward, Todd, Prelog, Lederberg, and Calvin all became Nobel Laureats subsequent to the date of this photograph, whether it be 1955 or 1961.

The first I met him in person was in the autumn of 1954 when we were both discharging not over-arduous duties of visiting professors, he at Harvard and I at MIT. We saw quite a lot of one another during that period and I can recall not infrequent triangular and somewhat convivial sessions with our mutual friend Bob Woodward. Since that time I have been privileged to remain in touch with him as a valued friend.—From an article entitled "William Summer Johnson: An Appreciation .... Lord Todd O.M., F.R.S." in *Bioorganic Chemistry*, 7, 121-123 (1978). (This issue no. 7 was dedicated to me in recognition of my 65th birthday, thanks to the editor, Gene van Tamelen.)

It was customary for the three of us to make the messages as obscure as possible as, for example, in the case of the letter written by Bob and Alex in 1956 (*see* Figure 1). Knowing of my aversion to acronyms, they particularly enjoyed ribbing me about the unusual number that crept into a publication on new reactions of diazomethane emanating from a joint effort at Caltech and Wisconsin. I countered by answering Bob in a relatively little known shorthand (Dewey), supplied by Barbara, which Alex recently admitted they had never been able to translate.

NMR Spectroscopy

The advent of the NMR spectrometer was probably the most important technical event in the history of structural organic chemistry. This methodology completely changed the *modus operandi* for characterizing the burgeoning numbers of new structures being produced by chemists involved in organic synthesis. My introduction to NMR spectroscopy was through J. D. "Jack" Roberts, who was the foremost pioneer in the use of this facility for the determination of the structures of complex organic molecules. (Jack has recently published his autobiography, *The Right Place at the Right Time*, within this Profiles, Pathways, and Dreams series.)

I knew of Jack Roberts by reputation but I had never met him until July 22, 1957, when he came to Wisconsin for a 12-day

Harvard University   DEPARTMENT OF CHEMISTRY

R. B. WOODWARD · 12 OXFORD STREET · CAMBRIDGE 38 · MASSACHUSETTS

April 28, 1959

Professor William S. ["Alphabet Soup"] Johnson
Department of Chemistry
University of Wisconsin
Madison 6, Wisconsin

Dear B:

Alex Todd and I were just sitting around recently, knocking back a few QO's, and he happened to mention your work on DM. We decided that I should W and congratulate U upon a brilliant contribution to the Sn of a problem which has bothered older OC's for generations. You will recall that Hermann Kolbe objected to the structural T on the ground that it made it possible for younger, inexperienced P to discuss OC with almost as much SF* as old dogs like himself. Your ingenious gambit will do much to redress the B.

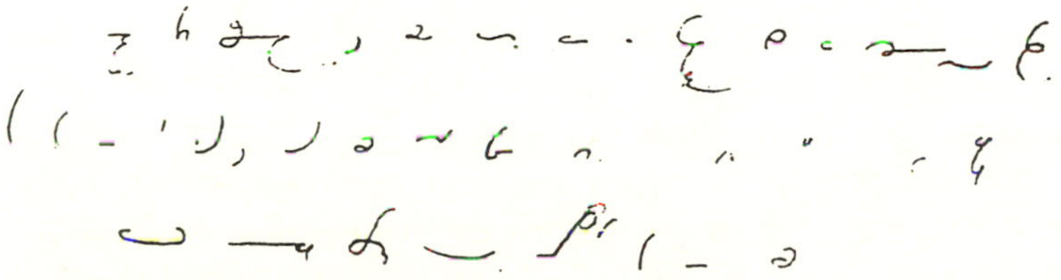

We found ourselves wondering whether the persistent rumors about the possibility of your going to SU involve your joining the OCD or the BCD.

Yours for obfuscation,

R. B. Woodward

RBW:DD
Copy to PSARTFRS

*In this paper, not to be confused with sang froid.

Figure 1. This letter, written a few years after the playful photograph of R. B. Woodward and W. S. Johnson jousting in the former's Harvard Office (page 70), further demonstrates their jesting. A key for a glossary of the abbreviations is found in Appendix I.

visit to deliver the Folkers Lectures. This visit was most memorable for me professionally as well as personally. A warm friendship began and then flourished leading to our introduction to Edith and the children, and to our now traditional (for over 27 years) family gatherings for Thanksgiving, to sailing expeditions, to tennis matches, and to other meetings—such events accompanied by much laughter and shouting as well as some wailing.

Soon after his arrival I asked Jack if he would show me how to operate our new NMR spectrometer. Jack agreed to take me on as a student, and I had the privilege of spending a good share of the 12 days with him because we became involved in other matters also (*see* subsequent section). A detailed account of this occasion may be found in a piece I wrote in 1969 entitled "My First Encounter with John D. Roberts."[60]

For almost two weeks I was busy day and night. Jack was in my office early each morning to find out what I had accomplished and to quiz me. He is a dynamic man of action with bulldog tenacity, who is not satisfied with anything much short of perfection. Those who remember the early 40 McVarian Associates V-4300-B spectrometer realize that the matter of obtaining satisfactory spectra in those days was really tricky. You had to learn how to recycle the magnet and then to search with the probe, sometimes for long periods of time, for a position where the field had some reasonable stability. Finally, after making numerous routine adjustments while scanning on the oscilloscope, you could start serious runs. With luck two or three satisfactory spectra might be obtained without interruption, but more often than not, in the middle of an early scan, someone would let the door to the NMR room slam, or there would be a power fluctuation, or a truck would go by, and the spectrometer would give out mostly noise; you were back at the beginning again. In such instances, long after a normal person was ready to give up for the day, Jack would show extraordinary patience, repeatedly coaxing the instrument into performance.

Once having learned how to cope with the instrument I worked night and day obtaining spectra on all sorts of research specimens, and I found myself in a state of euphoria because with Jack's help the revealing results poured forth. One example follows.

W. S. Johnson, Edith Roberts, and Jack Roberts at a party at Caltech (1970) celebrating the publication of the book entitled On Thirty Years of Teaching and Research *published by W. A. Benjamin, Inc. Among other things, this book contains a number of short stories told by friends of Jack. One of these is an account of my first exposure to Jack which is also described in the present book. (Photo courtesy of Andrew Streitwieser.)*

Among the NMR spectra we examined were those of the pair of crystalline ketols that had been already formulated as the two epimers **33** [A/B cis and A/B trans], because on treatment with base they each underwent elimination to give the same unsaturated ketone **34** (Scheme 10). The spectra of both ketols showed *two* high field sharp singlets of comparable intensity. (Note that there was no integration facility on that spectrometer.) One of these singlets had disappeared in the dehydration product **34**, so the one that remained obviously was attributable to the angular methyl group. Jack and I were poring over some spectra at my home after dinner on July 31. When our attention came around to the ketol our conversation went something like this:

Jack: Those peaks certainly look like two C–CH$_3$ groups.

Bill: But that's impossible; only one of them can be such a methyl.

Jack: Well, if one of those peaks isn't due to a methyl, it is the result of an exceptionally sharp, unsplit high-field methylene signal and I've never seen one like it. I still think it's a methyl.

Bill: But, there is just no way to formulate this ketol with two methyl groups attached to quaternary carbons.

Thus the conversation proceeded, until suddenly, by virtue of Jack's insistence, another possible structure for these ketols did occur to me, namely the bridged-ring structure (Scheme 15). This episode prompted a thorough reexamination of the chemistry of the ketols, which eventually proved that the structure (Scheme 15) was indeed correct. The base-catalyzed dehydration of the bridged-ring ketol to give **34** therefore proceeds via **33** in a reverse aldol process. Thus, it was demonstrated that the structure of all ketols from Robinson annulation reactions are open to question, and that the correct formulation is easily ascertained by NMR spectroscopy. Our detailed comprehensive paper on the nature of

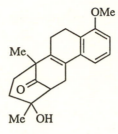

Scheme 15. Bridged-ring ketol from a Robinson annulation reaction.

the intermediary ketols in the Robinson annulation reaction[61] did not appear until much further experimental work was done, two and one-half years after Jack's memorable visit. However, a preliminary disclosure did appear on pp. 36–38 of Jack's well-known book *Nuclear Magnetic Resonance. Application to Organic Chemistry*, published early in 1959. I am the proud owner of a complimentary copy of his book, which carries the inscription "To Bill Johnson—my best pupil of NMR—Jack Roberts 1/7/59." Since then, Jack has had many pupils who are my superiors, but I doubt if he has ever had a more enthusiastic one.

Incidentals

Reactions of Diazomethane

My 12 days with Jack Roberts resulted in another research effort. On the morning of July 26, 1957, Jack was in my office as usual and we were discussing the matter of obtaining NMR spectra of some of the cevine alkaloids, which have numerous hydroxyl groups. I was deploring the fact that there was no really elegant way of exhaustively methylating all of the hydroxyl groups of such compounds, the classical method of prolonged treatment with dimethyl sulfate and sodium hydroxide being inapplicable because of the sensitivity of these substances to base. I submitted that what chemists really needed was a way of getting diazomethane to react readily with alcoholic groups in a manner similar to the way this reagent reacts with acidic hydroxyl groups to effect methylation. This comment reminded Jack of an observation he had made, that diazoacetic ester, under the influence of acid catalysts like p-toluenesulfonic acid, reacts with alcohols to form the alkoxyacetic esters and nitrogen.[62] Jack then proposed that diazomethane should methylate alcohols under acid catalysis provided an acid was used that would not react readily with the reagent. He then suggested that fluoroboric acid might work because it would produce a relatively long-lived diazonium cation that would react preferentially with an alcohol as the nucleophile rather than with the fluoroborate anion.

I was really impressed and said he should try out his hypothesis. Jack's reply was typical. In a very loud and firm tone that one might use to a misbehaving child, he said, "What are we waiting for? Let's do it now!" Thus, in a typical display of generosity, he invited me to share in the fruits of his idea.

So on the afternoon of July 26, having located some fluoroboric acid, we ran two experiments that are recorded in my laboratory notebook, No. 5, in my handwriting. One of these involved the gradual addition of an ethereal solution of a large excess of diazomethane to a solution of 2 mL of n-octanol in 15 mL of ether containing 2 drops of 48–50% aqueous fluoroboric acid. The addition was started at 3:25 p.m., and immediate evolution of nitrogen was noted. The addition was completed in 2 min, and the yellow color from the excess diazomethane persisted. After an additional 2 min, 2 more drops of catalyst were added, a vigorous evolution of nitrogen ensued, and the mixture became colorless. After treatment with anhydrous potassium carbonate, the supernatant liquid was analyzed on Harlan Goering's vapor phase chromatograph. The results are recorded in my notebook in Jack's handwriting: namely, that there was no peak corresponding to the original alcohol. Instead there was a shorter retention time peak that we assumed was due to the methyl ether, and the hoped-for conversion appeared to have occurred completely. The other experiment was with t-butyl alcohol, which also appeared to give the methyl ether, but the reaction was not complete.

As a consequence of these preliminary exploratory experiments, Jack and I laid plans for putting the method on a sound basis: He and Majorie Caserio in his laboratory examined the reaction with low molecular weight alcohols, and I with

Moshe Neeman at Wisconsin looked at the behavior of several steroidal alcohols. Out of this study came a joint communication to the editor on the methylation of alcohols with diazomethane[63] followed later by a detailed paper.[64] The conversion of cholesterol into its methyl ether in 95% yield by this method was later described in *Organic Syntheses* Vol. 41, 9, (1961).

In the course of applying the new methylation method to testosterone to give the then unknown methyl ether, Moshe Neeman noted the formation of a by-product that had lost the α,β-unsaturated ketonic function of testosterone. This observation was exciting because it had been well known that this type of steroidal α,β-unsaturated ketone system was inert to diazomethane. Thus the acid-catalyzed diazomethane ring enlargement of α,β-unsaturated ketones was discovered.[65,66] An impressive number of significant results developed from the idea Jack had in my office on the morning of July 26, 1957.

Constitution of Cevine and a Conformational Abnormality

While at Harvard in 1954–1955, I came to know Morris Kupchan who was working there as a National Institutes of Health (NIH) postdoctoral fellow. He was not happy because of lack of interaction with people interested in his studies in the cevine alkaloids; therefore, he asked if I would serve as his sponsor for continuing his fellowship at Wisconsin. Thus Morris joined my research group, and I took interest in listening to him describe his progress. He was an expert in the field and perfectly capable of running his own program. My input was minimal except in the following instance.

In a classical paper on the constitution of the hexacyclic polyterpenoid alkaloid, cevine, Barton, Jeger, Prelog, and Woodward[67] had assigned the configuration of the OH at C-16 as α (equatorial) because the acetate was extraordinarily susceptible to base-catalyzed methanolysis (*see* Scheme 16). Morris wanted to make the 16-β-OH (axial) epimer; so, at my suggestion, he hydrogenated the 16-keto compound over platinum oxide. He announced to me that this reaction, however, gave back the original "equatorial" isomer. Examination of the molecular model indicated that there would be serious steric interference to β-side compared with α-side approach by the catalyst, and I kept insisting that the original assignment was probably wrong: that is, the C-16 OH was β (axial). Morris found it difficult to accept my view

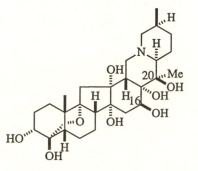

Scheme 16. Revised structure of cevine.

Vlado Prelog, W. S. Johnson, P. Plattner, Switzerland, February 15, 1951, "when I was an American–Swiss Foundation Lecturer."

in the face of the opinion of an array of supermen; however, at the same time he was excited with the idea that he might show that they were wrong. Further work did indeed substantiate the β-(axial) configuration of the C-16 OH, which in turn led to evidence for assignment of the β-configuration of the C-20 OH.[68-70]

More importantly, these studies revealed a new phenomenon, namely the powerful assistance of solvolysis of an axial acetoxy group by a neighboring hydroxyl in the 1,3-diaxial relationship. We concluded our first communication[68] with the statement: "One is forced to the conclusion that the simple rules of conformational analysis must be regarded with caution in treating complex structures." The 1,3-diaxial assistance of ester solvolysis was simultaneously disclosed by Henbest[71], and I have chosen to call the phenomenon "the Henbest–Kupchan effect".[72]

Thirty-Five Years at Stanford

During 1952–1956 I served on the American Chemical Society's Committee on Professional Training.[73] During a meeting in 1955, Art Cope pointed out that I was the only member of that group who had not held an academic administrative position. Knowing that the chairmanship at Wisconsin was about to become vacant due to Farrington Daniels' imminent retirement, and presuming that I was the most likely successor, he said that he was willing to bet that within five years I would be an administrator. Because I had no interest at all in such a job, particularly at Wisconsin where the faculty made the decisions and the Chairman had little authority, I simply told Art to name his stakes; the wager was a case of Scotch. Immediately some other members of the Committee wanted in on the action, with the result that I made the same bet with Louis Hammett and Bill Young. A year or so later when I refused to be nominated for the Wisconsin post, these gentlemen were resigned to losing their bets. However, because of some completely unforeseen developments that are set forth subsequently, they won and I paid off three cases of Scotch.

Nine-Year Involvement with Administration

In late 1958 I received a phone call from Fred Terman, the provost of Stanford University, saying that Stanford wanted me to become head of the chemistry department to spearhead their plans to upgrade it. I replied that I had no desire at all to become an administrator, that my primary interest was in research and teaching, that I had an ideal setup at Wisconsin with superb facilities, and a group of excellent students and postdoctoral associates, and finally that I had no inclination to leave. Indeed, we had just built a lovely new house on a beautiful lot in the Madison Arboretum, and we were particularly enjoying life then because my nephew, Tom Spencer (now Professor of Chemistry at Dartmouth) and his delightful wife, Patty, were on the scene, while he was working for his Ph.D. under Gene van Tamelen. Terman was not one to give up easily, and asked if Barbara and I would be willing to pay them a visit with the purpose of at least giving them advice. As it was winter in Madison, the idea of the trip was not unappealing; so we went for the visit.

To make a long story short, I was immensely impressed with President Wallace Sterling and Fred Terman who ran the university in a benevolently despotic manner, making all major policy decisions and approving all appointments, with minimal faculty involvement. In a crisis, faculty input could be effective. For example, when the Stanford Linear Accelerator (SLAC) program received Terman's approval to have a Ph.D. program of their own, thus competing with the Physics Department, a group of us went to Sterling to plead Physics' case, and he overrode Terman's action. Most of the departments were similarly run by executive heads who were expected to make unilateral decisions. Many of the faculty were unhappy with the system. George Pake, then in the Physics Department, was one of those who felt the faculty should have much more to say in running the University. I used to argue with him, pointing out that he should enjoy being in a well-run university and not having to bother with slow, time-consuming meetings of committees, councils, senates, and the like. Ironically, after becoming provost of Washington University (St. Louis), George told me he had come around to my view. In the meantime, he had been instrumental in Stanford's gradual changeover to a highly democratic system, which of course, is a protective device against incompetent top administrators.

Be that as it may, Stanford's early 1960s administration was ideal for effecting rapid, dramatic changes; indeed, several departments such as Biochemistry, Mathematics, and Physics had already become great under this regime. Sterling and Terman assured me that it was now Chemistry's turn. I also found that the department, already consisting of a number of talented scholars, was highly supportive of the new plan; hence I returned to Wisconsin with the strong feeling that Stanford's aspirations were bound to be realized, and I was somewhat inclined to become involved. My enthusiasm increased after I had the idea of finding a top organic chemist to join me in the proposed move. Carl Djerassi, then at Wayne State, was my choice. In a memorable conversation he and I had on the matter at O'Hare Airport, he caught my enthusiasm for the Stanford plan and as we parted Carl said, "I think you and I will probably end up at Stanford." Terman was delighted with this development and immediately began negotiations with Carl.

On April 15, 1959, I wrote a five-page acceptance letter to Terman with numerous contingencies relating to funds needed for new faculty appointments, major laboratory facilities such as NMR spectroscopy, remodeling of the lecture halls, etc. This letter was written with Carl's help and approval so that it also represented an acceptance in his behalf. Somewhat to my surprise, Terman immediately replied that Stanford agreed to all of the contingencies.

Leaving Wisconsin and our friends there after 19 very good years was difficult, particularly for Barbara. Nevertheless, we adjusted quickly to the new environment. After Carl's and my acceptance, Sterling and Terman forthwith raised funds for a new building to accommodate our programs. The donors were members of the family of John Stauffer, after whom the building was named. Carl and I were deeply involved in the design of the building.

Carl's group came on before completion of the building and in August of 1960, I moved directly into the new facilities. Research commenced immediately, as I had brought a dozen seasoned collaborators with me from Wisconsin. One of these, Ken Williamson, is unique in that he worked with me at three different institutions (B.A. research, Harvard 1955; Ph.D., Wisconsin 1960; head postdoc., Stanford 1960–1961). Ken and I first became involved when he was an undergraduate at Harvard and I was a visiting professor. He had chosen me to be his mentor for a research project. The upshot of this relationship was that Ken came to Wisconsin for graduate work where he chose me as his major professor. He performed so well that when he completed his work

"The Three Musketeers." Gilbert Stork, Carl Djerassi, and W. S. Johnson, 1976. (Photo courtesy of Abhijit Mitra.)

This shot of Barbara and me was taken on the occasion of a farewell party given to us by the Chemistry Department (Madison, 1960).   The trumpeter is the late Charles Heidelberger of fluorouracil fame who then was professor of oncology at Wisconsin. Charlie used to have jam sessions at his home and I occasionally participated, playing tenor sax. (Photo courtesy of J. Doplf Bass.)

for the Ph.D. in 1960, the year I was moving to Stanford, I invited him to assume the position of head postdoctoral assistant and to be in charge of moving a large group of co-workers and equipment. Thus he remained with me until 1961. Ordinarily I would not approve of nor promote such a one-sided sort of training; however, Ken does not seem to have suffered from it; indeed he made a fine record on his own, particularly in the field of NMR spectroscopy.

Movement of academics from one university to another is not infrequent. The poem shown in Figure 2 illustrates the humor of scientists while talking about Carl Djerassi's and my move to Stanford and E. J. Corey's move from Illinois to Harvard. It also pokes fun at Carl's propensity for publication, a subject that Carl himself talks about with some humor in his own autobiography within this Profiles, Pathways, and Dreams series, *Steroids Made It Possible*. The identity of the author is not known. E. J. wrote to me

W. S. Johnson, John Stauffer, Jr., and Carl Djerassi, at the dedication of the Stauffer Chemistry Building, Stanford University, 1961.

on December 1, 1993, "With regard to the poem, I had seen it years ago, having been sent a copy. I believe that Bob Woodward also received a copy. I am sorry that I do not know the identity of the author. Ralph Raphael is a possibility, but it is certainly not his style."

One of the stipulations of my acceptance of the department head at Stanford was that I should have an associate head who would inhabit the main office and handle day-to-day administrative matters, while I attended to policy matters, teaching assignments, salaries, and faculty promotions and recruiting. As it

Kenneth Williamson in the lab, 1961.

It has been said that "....the cream of the Middlewest
is being extracted by ethereal overtures from the coastal
states." In the light of this saddening fact the memorial
poem which appears below seems not inappropriate.

My name is Bill Yonson,
I came from Visconsin--
I worked on de steroids dere;
De people out Vest
Dey say it's de best,
So I give ol' Visconsin de air.

My name is Djerassi,
My structures are classy--
I worked on the Mexican yam;
Rotations infernal
On terpenes eternal
In every month's journal I cram.

My name's E. J. Corey,
Success is my story--
Next year at Harvard I reign;
Though down in Urbana
I was top banana,
I only liked bottled Champaign.

Anonymous

Figure 2. This poem was written sometime in the late 50's to commemorate the "translation" of Carl Djerassi and W. S. Johnson to Stanford and E. J. Corey to Harvard from Illinois. It's authorship remains unknown. On December 1, 1993, Corey responded to Johnson's request for information as follows: "With regard to the poem, I had seen it years ago, having been sent a copy. I believe that Bob Woodward also received a copy. I am sorry that I do not know the identity of the author. Ralph Raphael is a possibility, but it is certainly not his style." Should anyone be able to identify unambiguously the author, please be encouraged to contact the Editor.

turned out, this arrangement left time for research with a group of about 20 collaborators (*see* subsequent section).

Faculty recruiting was my most important administrative job. Under the autonomous system I was expected to select the targets, orchestrate the wooing procedures, and perform all negotiations. The job was made relatively easy in that Terman arranged for me to bypass the deans and to report directly to him. This facility for acting rapidly and decisively was very efficient and proved to be essential in making one of our most important appointments, namely that of Paul Flory, as shown in the following account. In December 1960, when I was serving on a National Science Foundation (NSF) Fellowship Committee in Washington, D.C., I heard another member of the committee, Al Blomquist, mention that Paul Flory was resigning as Director of Mellon Institute to take an academic position. Al, who was then Chairman of the Chemistry Department at Cornell, said that he was expecting Paul to accept their offer. As soon as I returned to Stanford, I contacted Terman and within 15 minutes had approval to make Flory an offer. When I phoned Paul, whom I had never met, he indicated that it was probably too late to become involved, because he was committed to reach a decision soon regarding three other academic offers. However, I adopted Terman's strategy and asked him if he would at least pay us a visit and he obliged. During his visit the weather was uncooperative providing us with a typical windy winter rainstorm. In response to my apologies, Paul wrote:

> Incidentally, any adverse prejudice which might have been engendered by the sample of California weather over the weekend was dispelled immediately upon my return to Pittsburgh in the midst of a raging blizzard. The plane was delayed on this account, driving to my home area was slow and hazardous and, to cap it all, I could not get up the hill a mile from my home. Taxis had mysteriously disappeared. Finally, I sought mercy from the local constabulary, who kindly took me home. The hour was late even on California time. When next I am brought home by police car, my wife will insist upon some other explanation.

In any case, Paul had become intrigued with the Stanford scene for the same reasons that I had. This visit was followed soon by another that included his wife Emily. His acceptance on February 6, 1961, had a profound effect on our program. Henry Taube (then at Chicago), whom we had been trying to attract, soon decided in our favor. In a biographical memoir on Flory, Henry wrote, "Flory, with characteristic decisiveness, made up his mind before I did, and his decision made in 1961 to accept an offer from Stanford influenced my own. By then his scientific reputation was widely and firmly established, and by then I had met him and his wife Emily several times. All factors contributed as strong inducements to join him as a colleague."

The reputations of Flory and Taube, both of whom were destined to receive the Nobel Prize, placed them, even in the early 1960s, among the few truly distinguished scientists. Hence with them on the roster, it became relatively easy to attract top scholars like Gene van Tamelen from Wisconsin in 1962 and Harden McConnell from Caltech in 1964. These early appointments were responsible for the spectacular rise of Stanford's Chemistry Department from 15th position (in 1957) to 5th (in 1964) in the nation, according to the 1966 report of the American Council on Education, "An Assessment of Quality in Graduate Education". One matter that concerned me from the beginning was the question of how long it would take before Stanford would attract the top graduate students. Gratifyingly, the response was almost immediate. Since then, when I have been asked for advice on how to attract top students, I have said, "Add some distinguished scholars to your faculty." Fancy brochures are, in my opinion, ineffective.

It should be emphasized that we certainly did not have a perfect recruiting record by any means. Thus considerable effort went into the unsuccessful wooing of a number of people including, among others, Derek Barton, Ron Breslow, and Manfred Eigen. These three showed particular interest, but not quite enough to make the move.

Another important appointment, that I luckily made, was the promotion of Doug Skoog (already a full professor in the department) to associate head. He performed superbly in this capacity and continued to so serve the department for many years after I gave up the administrative post. It is hard to overemphasize how important it was to me to have the strong support of the pre-1960 faculty, particularly the organic group consisting of Bill Bonner, Dick Eastman, Carl Noller, and Harry Mosher. I have had the pleasure of being involved with Harry, not only as a colleague, but quite regularly on the tennis courts for many years. Harry's game is very good and I rarely took a set from him.

In addition, I was blessed with magnificent secretarial help: Anne Foley, 1960–1967; Marlene Bunch, 1968–1970; and Carolyn Southern, 1970–present. After 25 years, Carolyn has become such an important part of the operation that it is inconceivable to imagine doing without her.

From the beginning I always consulted with the senior faculty for approval of most of the major issues. Mainly with the aim of taking advantage of Flory's wisdom and expertise in administration, I established an Executive Committee (the first one being composed of Flory, Skoog, and myself as chairman) with the charge of addressing issues raised by the chairman, and tak-

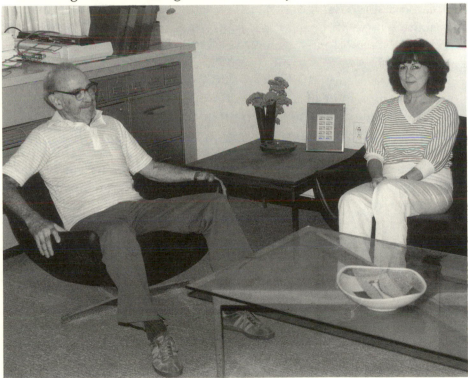

Carolyn Southern, my loyal secretary since 1970.

ing action on many matters that otherwise might be brought before the whole department. This practice was so successful that
this system has prevailed in the department to this day.

Within a decade nine established scholars had been added
to the department. In addition to those senior faculty named
above, John Baldeschweiler and Jim Collman both joined the department in 1967, followed by Linus Pauling and Kenneth Pitzer
in 1969. Aside from recruiting established chemists it is, of course,
important to keep bringing in junior faculty members in the hope
of finding an occasional star via this route. During my nine-year
term as department head we engaged nine assistant professors,
three of whom, namely Hans Christian Andersen, John Brauman,
and Robert Pecora, were given tenure and have since been widely
recognized as distinguished scholars and have developed into
important leaders in their fields.

Courtesy professorships were extended to Michel Boudart
(1964) and Karl Folkers (1964–1969). When Chemical Engineering
became a separate department in 1960, David Mason was made
executive head and has held a joint appointment with Chemistry.
The Masons and Johnsons have been especially close friends from
the beginning.

Now I should like to record some retrospective observations on the matter of trying to improve a chemistry department.
In 1959 the major research strength of the Stanford Chemistry Department was in the organic area represented by William (Bill)
Bonner, Richard (Dick) Eastman, Harry Mosher, and Carl Noller.
With the addition of Carl Djerassi and myself the department was
heavily biased toward organic chemistry. Physical chemistry was
represented by Eric Hutchinson, Frederick Koenig, and the distinguished Richard Ogg who died in 1963. The obvious immediate objectives were to bring in some leading scholars in physical
and in "modern" inorganic chemistry, which was completely un-
represented at Stanford. At the same time, I was inclined to add
one or two more distinguished organic scholars with the aim of
immediately making Stanford one of the important places for this
field. My nonorganic colleagues were not particularly enthusiastic
about this plan until Terman gave it his blessing. He said that in
endeavoring to develop a great department a good way to start is

The late David Mason, who was largely responsible for transforming the Stanford Chemical Engineering Department into a world-class operation. David was a special friend of mine. He and I are shown here in a familiar position, playing jazz duets.

by erecting one high spire that can be seen all over the world and then gradually build around this. Thus in 1962 we were able to lure to Stanford my former Wisconsin colleague, Gene van Tamelen, who at the tender age of 35 was already widely recognized as a scholar in a class with Woodward and Corey.

During our efforts to bring distinguished scholars to Stanford, I learned that in the 1960s when granting agencies were inclined to be generous in supporting established scholars, financial support was not the most important factor. The candidate must feel that he is moving to a place that has or will have a better reputation and can attract better students than the place that he or she would be leaving. Preferably the candidate would be leaving a place where he or she has lived at least 10 years and a change is appealing. In other words a push as well as a pull is generally needed.

It is not difficult to develop a list of distinguished scholars that would make your place number one; the problem is to get them to move. If they have tenure at Harvard, it's almost certainly a lost cause. If they are located elsewhere, there is a chance depending on his or her length of tenure (10 years is apt to be a minimum of time for minor to become major irritations, thus making change to a set of new minor irritations more attractive). If the atmosphere of your university is attractive physically as well as intellectually a visiting professorship appointment can be helpful, as it was in the case of Henry Taube (i.e., Palo Alto vs. Chicago). The recruiter must be prepared to face disappointments. Thus, Manfred Eigen, who spent a quarter with us as visiting professor, finally, and with obvious reluctance, turned down our offer to come to Stanford permanently. Others who seemingly came close to accepting our offer were Ronald Breslow and Derek Barton. In the Barton case the timing was not right.

Next to Nobel Prizes, Fred Terman used to count the number of members of the National Academy of Sciences as an evaluation of science departments. By this measure the "new" program at Stanford did well: in 1959 the number was 0; in 1961, 2; 1962, 4; 1964, 5; 1969, 7. Terman was so pleased with the rapid way in which the new program was developing that I could easily get his support for almost anything we wanted. Terman's strong prejudice in my favor is illustrated by a letter he wrote on July 14, 1970, congratulating me on receiving the (French) Roussel Prize. As I said in my reply, he (Terman) had a grossly exaggerated opinion of the role I played in the development of the new department. It was a joint effort in which many people were involved, particularly Terman himself. Moreover, we were also very lucky and happened to be at the right place at the right time as, for example, in the case of Paul Flory.

During the 1960s the affairs of the department were being handled in an increasingly democratic fashion. When I resigned the headship in 1969 Terman recommended that the department

July 14, 1970

Professor William S. Johnson
Department of Chemistry
Stanford University
Stanford, California 94305

Dear Bill:

Congratulations on the attached! The honor has special significance, since as the initial recipient of this prize, there must have been a healthy backlog of outstanding candidates in the competition.

However, in spite of the honors that you keep getting from your professional colleagues, to me your greatest achievement has been the Chemistry Department that you have created at Stanford. In this conection, you will be pleased to know that when the American Council on Education publishes its new ratings of faculty quality this fall (based on spring 1969 evaluations), Chemistry at Stanford will have edged significantly ahead of the fifth position received in the 1966 publication (which gave ratings based on spring 1964 evaluations). The new information is confidential, but nonetheless authentic.

The transformation that you have achieved in Chemistry at Stanford is probably without parallel in the history of education in Chemistry, since it was done with only a modest injection of new funds, and without producing a swollen oversized Chemistry Department. You certainly exploited the basic principle understood by so few that the quality of an educational program depends much more on the person on whom one spends his money, than on how much money is spent, or the gross number of new appointments made!

Very truly yours,

s/Fred
F. E. Terman
Provost Emeritus

was ready to run itself: that is, in a democratic fashion with a three-year term chairman. Once a department reaches a certain level of excellence, it is self-perpetuating. Thus under the chairmanship of Flory, van Tamelen, Taube, Brauman, Ross, and McConnell the Stanford department has thrived so that by 1993 it enjoyed the distinction of having three Nobel Laureates and 12 members of the National Academy of Sciences among other honors. It is a pity that Fred Terman, who died in 1982, is not able to enjoy these results of his efforts.

The Hydrochrysene Approach to Steroids, Continued

Refinement of the aldosterone synthesis was a major effort that comprised the thesis of Paul Kropp (Ph.D. 1961). I was then confronted with the task of writing the full paper (no. XV of the series), which was the longest one we had ever prepared for *JACS*, covering 21 pages in the Journal.[51] Also in the early Stanford days, John Keana (Ph.D. 1964) synthesized specimens of racemic cholesterol[46,47] and progesterone.[47,49] These were challenging exercises in elaboration of the established totally synthetic nucleus and our publications represented the first report of a completely total synthesis of these important substances.

When I was telling Bob Woodward about this work before it was published, he mentioned that one of his collaborators had carried his steroid synthesis all the way through to progesterone by using racemic materials (i.e., without the use of relays), and when the last step was reached, chromatography yielded optically active progesterone! Not surprisingly, this startling result could not be duplicated after repeated attempts. Bob felt very uncertain about this work, which he therefore never published. He provided us with a specimen of his material, which John Keana found to melt over a wide range. When the Harvard material was recrystallized, the mp rose to 183–185 °C and showed no depression on admixture with our specimen, mp 183.5–185.5 °C. Moreover, the IR spectra of the two samples were identical. Therefore, Woodward had indeed prepared the racemic hormone and we mentioned this fact in our paper, although no other information is available on his synthesis.

Conessine

In 1960 one of my academic grandchildren, Jim Marshall (Ph.D. with Bob Ireland) started a two-year NIH postdoctoral stint with me. I sold him on a plan directed toward conessine by using a strategy similar to that already worked out for aldosterone (Scheme 14), but without the oxygen atom at C-11. Thus the angular carboxy group of the compound at stage **45** was envisaged as becoming the N-substituted aminomethyl bridge of conessine (Scheme 17). Manipulation of the tricarboxylic acid system was giving some real problems, and one day Jim came into my office

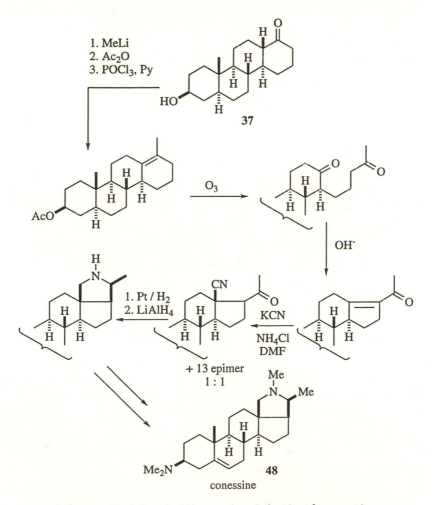

Scheme 17. Marshall's total synthesis of conessine.

with a new, imaginative plan that was quite rational and relatively short; however, it required starting all the way back with
substance **37**. Therefore, I submitted that the original plan was already far along and was a bird in the hand. We had a fairly long,
friendly argument about this matter and finally I gave in to him,
mainly because I have always liked to have my collaborators
work on their own ideas, if they are good ones, because it provides an added stimulus to success.

In Jim's case there was the proviso that if he got into real
trouble with his own route (and I pointed out some places where
this might occur), he would return to my plan. As it turned out,
Jim, with great determination and skill, made the scheme shown
in Scheme 17 work very nicely indeed. His synthesis is shorter
and provides more novel strategy than was involved in my plan,
and he is the principal author of the communication.[48] On the occasion of completion of the synthesis Jim, his wife Anne, Barbara,
and I had a memorable celebration at our home.

Veratrum Alkaloids

The basic chemistry for contracting ring C of the tetracyclic substance **40** was developed by Peter Schiess (postdoc. 1960–1961).[52]
This methodology was applied to the total synthesis of formula
53, Scheme 18, via the steps **49** → **50** → **51** → **52**.[54] Substance **53**
had been prepared from veratramine by an ingenious degradation
sequence developed by Dick Frank (Ph.D. 1962).[74] When Dick
started graduate work he honestly admitted that he was not at all
sure that he belonged in chemistry, but he intended to give it a
good try. About the time he discovered the interesting fragmen-
tation reaction that led to the aforementioned degradation of
veratramine, he became totally turned on by organic chemistry
and has remained so to the present day, a most gratifying turn of
events for me as well as him. Aldehyde **53**, being readily available, served as a relay for the formal total synthesis of veratramine, which was skillfully completed, in 12 steps, by the joint
crash effort of four of my collaborators.[53]

Another person who made significant contributions to the
synthesis of the tetracyclic ketone **52** was Noal Cohen (postdoc.

1965–1967), who was one of my academic grandchildren (Ph.D. with Marshall). It was not until Noal was near the end of his work at Stanford that I learned that he is a highly talented jazz musician. His specialty is drums, which he had played regularly with

W. S. Johnson and Jim Marshall, conferring on the conessine project, Stanford, 1962. "The huge bottle of 'd,l-8-isoconessine' is a joke." According to Marshall, Bill and I were "fabricating an enormous jar of what could be assumed by its large label to be synthetic 'd,l-8-isoconessine' (actually the jar contained salt). The implication of this ploy was that the synthesis of 8-isoconessine [a key intermediate in Gilbert Stork's concurrent conessine synthesis] was a trivial matter compared to the real stuff [which we had synthesized]. We planned to maneuver this jar into the C&E News photo so that it would be noticed by Stork and his co-workers when the article appeared. The Rube Goldberg array of pipes on the Erlenmeyer flask to the right of the 'd,l-8-isoconessine' is a gag vermouth dispenser donated by WSJ for the photo op."

Chuck Mangione. Also he had played on occasion with greats like
Oscar Peterson, and personally knows many of the top artists.
Needless to say I was thrilled to discover this facet of Noal, who is
very modest and had given up music, at least temporarily, in the
interest of chemistry. Since he left Stanford we have kept in touch,
meeting when the opportunity presented itself to talk about, listen
to, and even play a little music.

Concluding Remarks

The major significance of the hydrochrysene approach to steroid
synthesis was that it represented an early example of the possibility of designing syntheses of complex molecules in such a way
that a multiplicity of asymmetric centers could be introduced
stereoselectively. This study therefore was a factor in a major
change in attitude regarding steroid synthesis from the two- to
the three-dimensional approach.

When one's methodology is used by others, this represents
a true compliment. Thus Nagata et al. used the method shown in
Scheme 10: starting with 2,6-dimethoxynaphthalene they synthesized an isomer of **34** with the methoxyl group in the 2 instead of
1 position, which was converted into steroids by novel elaboration of the D-ring.[75] It is noteworthy that Willy Bartmann
(postdoc. 1958–1959) synthesized this same isomer of **34** with
similar objectives to Nagata's in mind; however, I was preoccu-
pied with the impending move to Stanford and did not write it up
for publication in time. My apologies to Willy. In some other
studies, Kutney et al.[76] adapted the Schiess ring-C contraction
methodology, illustrated in Scheme 18, to Nagata's isomer of **34**,
and elaborated the D-ring to produce veratrum alkaloids.[77]

Biomimetic Polyene Cyclization Studies

By 1960 it had been rigorously established by Bloch and Woodward[59] that the key step in the biogenesis of steroids was the sensational conversion of the acyclic polyene, squalene, into the tetracyclic triterpenoid, lanosterol, involving the formation of seven

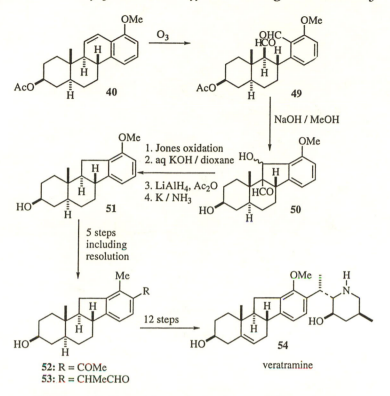

Scheme 18. Formal total synthesis of veratramine.

asymmetric centers. Also, the ideas of Stork (*see* the photos of Gilbert on pages 62 and 64) and of Eschenmoser relating to the mechanism of squalene cyclization had been announced.[78,79] They postulated that the process is a cationic concerted olefinic polycyclization that is stereospecific, involving *trans*-coplanar additions, so that *E*-olefinic bonds result in *trans*-fused rings, whereas *Z*-olefinic bonds give *cis*-fused rings. An illustration of how the hypothesis applies to a cyclization of squalene is depicted in Scheme 19. By 1960, it had finally been established that nonenzymic, acid-catalyzed cyclization of polyenes did not give promising results.[80] Bicyclic material was produced from a trienic substrate in 60–70% yield; however, in opposition to theory the configuration at the ring fusion was the same irrespective of whether the internal bond was *E* or *Z*. Moreover, a tetraenic substrate gave only a 5–10% yield of all trans tricyclic product, in regard to which Albert

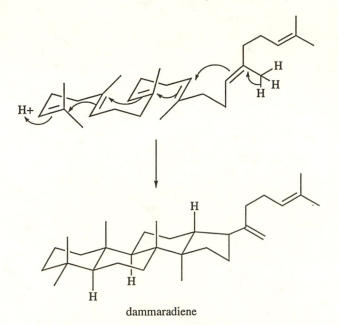

dammaradiene

Scheme 19. Illustration of the Stork–Eschenmoser hypothesis.

Eschenmoser, who in 1959 was already recognized as a foremost leader in organic chemistry, wrote:[80] "Apart from it and from a similarly small amount of bicyclic dihydroxy compound, the main product consists of an intractable oily mixture." He concluded: "It seems that with polyenes of this complexity, acid-catalyzed cyclization ceases to be a useful reaction from the preparative point of view." Albert's conclusion was, and still is, entirely correct in the context of the types of systems under consideration, namely where olefinic bonds are the cyclization initiators.

　　　Thus the stage was set for taking a new approach to the problem, and we initiated a study that, after 35 years, is still in progress. We had no idea how far this project might go or that it would lead to spin-off studies of significance. We just followed our noses, with high hopes. The new strategy was based on the feeling that the use of the strongly acidic conditions that are required to initiate cyclization of polyenes is bound to give poor results, because protonation will be indiscriminate and will produce carbocations at unwanted sites. Hence, the idea was born of using

Some of the older participants in the Table Ronde Roussel UCLAF No. 30, March 3–4, 1977, titled "Computer-Assisted Design of Organic Synthesis." I can't remember having been photographed with a more illustrious as well as amusing group of chemists. Left to right: Vlado Prelog, R. B. Woodward, Albert Eschenmoser, myself and Jean-Marie Lehn.

a functional group at the end of the polyene chain to act as an initiator and, in principle, we anticipated the natural process involving 2,3-oxidosqualene, which was not discovered for several years after our studies were begun.

It is important to emphasize that the epoxy group was not among those that we had considered as candidates for cyclization initiators. Hence, when Gene van Tamelen, as well as E. J. Corey, disclosed its role in the enzymic process,[81] it was perfectly natural for Gene also to become involved in studying biomimetic polyene cyclizations. Contrary to the opinion held by some chemists, Gene and I were not competitors, although I think that some of our respective students felt that they were adversaries. As it turned out Gene chose to confine his studies mainly to the epoxy-initiated processes and I looked at other (less natural) functional groups. Thus the two programs were quite complementary, and we frequently exchanged information and ideas, which was a good protection against developing a conflict of interests. Also our objectives were somewhat different. He was concerned particularly with the matter of finding out what the differences were between the enzymic and nonenzymic processes, as well as learning what happens to epoxy squalene analogs in the enzymic system. These studies, many of which were summarized in review articles,[82] provided highly significant information regarding the enzyme mechanism.

My objectives, on the other hand, were directed mainly at developing methodology for the efficient production of polycyclic systems without close adherence to the biological system; specific targets, such as the synthesis of corticoids, were given much attention. In one of his more recent studies, however, Gene developed a very ingenious (unnatural) application of an epoxide initiated cyclization that resulted in a fairly efficient steroid synthesis.[83] Also, by chance, some of our recent studies have provided inferential documentation for a mechanism of enzyme action (*see* subsequent discussion).

Attention is now turned to a perspective of the whole program, from its evolution to the present status. Comprehensive summaries of our work through 1976 may be found in two review articles.[84,85] Also more recent overviews of the field have been made by Paul A. Bartlett[86] and by Groen and Zeelen.[87]

(Left to right) Gene van Tamelen, Barbara and myself, Mary van Tamelen, at the Caribbean Chemical Conference, University of the West Indies, Barbados, 1969. (Photo courtesy J. D. Roberts.)

Solvolysis of Sulfonate Esters

It had been established, particularly by the eminent physical organic chemist Paul D. Bartlett at Harvard, that the solvolysis of
hex-5-enyl sulfonate esters proceeds with rate enhancement involving participation of the olefinic bond to give cyclic products.[88]
We therefore undertook a study of the solvolysis of dienic sulfonate esters.[89] Although the yield of bicyclic materials from the
trans-ester (Scheme 20) was disappointingly low, complete analysis of the product mixture showed that all of the bicyclic materials
belonged to the *trans*-decalin series.[89a] Conversely the *cis* dienic

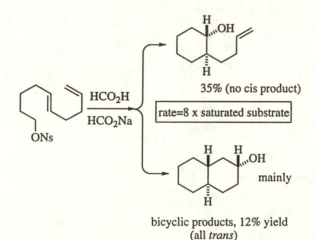

Scheme 20. Functional group initiator concept.

ester (Scheme 21) yielded only *cis*-decalin products,[89b] which is the
expected product from the postulated *trans*-coplanar addition
process. Thus, for the first time, strong documentation was provided for the Stork–Eschenmoser hypothesis in a nonenzymic
process.

It was not surprising that solvolysis of a trienic ester gave
a very poor yield of tricyclic product (Scheme 22).[89e] This was the
state of affairs when I reported on the subject in London at the
International Union of Pure and Applied Chemistry Conference
in 1963,[90] and the general feeling among the experts was that the

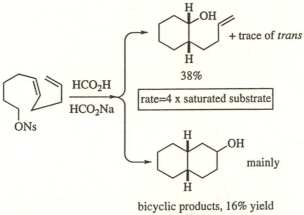

Scheme 21. Corollary evidence for the Stork–Eschenmoser hypothesis.

R. B. Kinnel, now at Hamilton College, was a postdoc with me in the mid-60s. Here he proudly stands by his "new used Pontiac" in Menlo Park, early 1966.

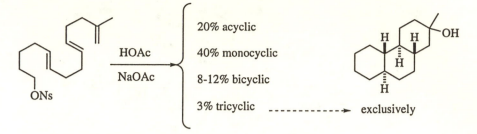

Scheme 22. Attempted tricyclization.

unfavorable entropy of activation for such processes precluded the possibility of realizing efficient nonenzymic polycyclizations. Nevertheless, we did continue some solvolysis studies because of their theoretical interest. In the aforementioned solvolysis studies of Bartlett,[88] it was demonstrated that the rates increased significantly as the double bond was substituted by groups that stabilize the cyclic cation (a first-order anchimeric effect). The rate data summarized in Scheme 23 indicate that in the process that produces bicyclic material there is a second-order anchimeric assistance by the remote olefinic bond, which may be regarded as evidence for a concerted cyclization. Part of the results are reported in the Ph.D. theses of Hal Doshan and Glenn Prestwich. Because of the untimely death of Dan Sargent in 1977, who performed some of the significant rate studies during his sabbatical leave from Amherst in 1969–1970, this work has not hitherto been published. His records were never retrieved from Amherst, although summary information is at hand.

It is estimated that about 26 human-years of work went into this sulfonate ester solvolysis program. Why should such work take so much time? As an illustration consider the case shown in Scheme 20. The gas chromatogram of the product showed 16 peaks; the products corresponding to 11 of these were separated and the structures and configurations were proved. It was particularly important to demonstrate whether the bicyclic substances (the minor ones were hydrocarbons with one double bond) belonged to the *cis*- or *trans*-decalin series before we could draw any conclusions about the bridgehead stereoselectivities of the ring closures. In the case of the rate studies it was essential to obtain information on the structure and yield of bicyclic material for each example in Scheme 23.

Solvolysis of p-nitrobenzenesulfonates

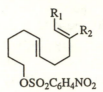

R_1	R_2	Rel. Rate[a]	% Bicyclic Product[a]	Rel. Rate[b]
H	H	1.00	9	1.00
H	CH_3	1.16	16	1.15
CH_3	H	1.25	20	1.36
CH_3	CH_3		25	1.46

[a]With H. Doshan (1967) and G.D. Prestwich (1974);
0.01 M substrate, 0.02 M NaOAc in HOAc

[b]With G.D. Sargent (1970);
0.01 M substrate, 0.04 M urea in CF_3CH_2OH / 3% H_2O

Scheme 23. Evidence for second-order anchimeric assistance in a bicyclization.

Some Success (in the Nick of Time)

The morale of my collaborators who were searching for potential initiator functions was about as low as the yields of our cyclizations, and I was seriously considering abandoning the whole study, when two different systems showed some real promise of giving high yields in bicyclizations. These involved the use of an acetal and of an allylic alcohol as initiators, illustrated by the examples shown in Schemes 24[91,92] and 25.[93,94] The configurations of the cyclization products in Scheme 25 were proved by their conversion to the natural product, fichtelite, of known constitution. It is extraordinarily significant that in the case shown in Scheme 25, the process is *not* interrupted after one ring is formed, even though a nucleophilic medium (formic acid) is employed. This result is in striking contrast to the sulfonate ester solvolysis

(Schemes 20 and 21), where the major product is monocyclic material produced by reaction of solvent before completion of the bicyclization. The difference in these two processes shows that the nature of the initiator is extremely important in the success of a polycyclization. The theoretical characteristics of a good initiator have been developed from extensive study. An important empirical rule emerged from early studies: namely, *unless an initiator gives nearly quantitative yields in bicyclization, it will not be satisfactory for promoting polycyclizations.*

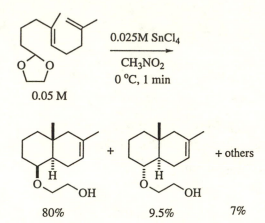

Scheme 24. Acetal initiator.

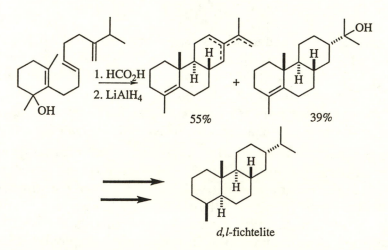

Scheme 25. Ditertiary allylic cation initiator.

A very successful bicyclization that was for the most part conceived and reduced to practice by Paul A. Bartlett (Ph.D. 1972) for his thesis is shown in Scheme 26.[95] Paul, incidentally, remains a good example of the kind of student for whose success a professor might unjustifiably try to take credit. The cyclization step (Scheme 26) was quantitative and so clean that Bob Volkmann, after attenuating the rate by using milder conditions, was able to perform kinetic studies monitoring product formation as well as disappearance of starting material. A Hammett plot of the data in Scheme 27 shows considerable curvature; so we appealed to my young colleague, John Brauman, for help in the interpretation. He suggested that a reversibly formed intermediate, probably complexation of the catalyst with substrate, was involved and that in the very rapid cyclizations (CH_3 and CH_3O elicit the same rate), the complexation was rate controlling. He worked out the mathematics for this postulate, and found the rate data to be quite compatible. In a joint publication[96] a convincing case for second-

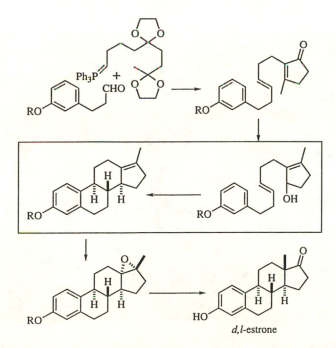

Scheme 26. Secondary–tertiary allylic cation initiator; Bartlett's stereoselective estrone synthesis.

order anchimeric assistance (*see* previous discussion) of the cyclization by the aromatic nucleus was made.

I digress briefly to mention that since Paul Bartlett obtained his Ph.D., he and I have continued to have close interac-

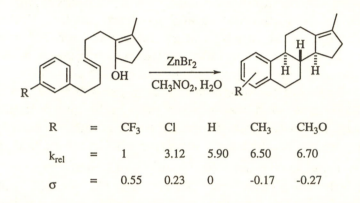

R	=	CF$_3$	Cl	H	CH$_3$	CH$_3$O
k$_{rel}$	=	1	3.12	5.90	6.50	6.70
σ	=	0.55	0.23	0	-0.17	-0.27

Scheme 27. Rate study showing evidence for a concerted bicyclization.

W. S. Johnson and Paul Bartlett at the Cope Award dinner banquet, Miami, 1989. (Photo courtesy of Andrew Streitwieser.)

tions to the present day. Although his own research interests have carried him to new areas, his concern about the field of polyene cyclizations has never waned. Indeed, he is one of the foremost authorities in the field and has written an important treatise on the subject,[86] which he dedicated to me. I am warmly touched by this tribute. As recently as 1982, Paul became involved in a joint venture with us that opened up a new area of study described at the end of this section. As the years have gone by a very warm friendship has developed between us. Thus Paul has regularly arranged some sort of a birthday celebration for me, an especially memorable one being a large party of my collaborators and other friends held, in 1979, on a boat that embarked from Oahu, Hawaii. Later, for my 70th birthday, Paul, along with Fred Li (postdoc. 1967–1969) and Mike Garst (postdoc. 1975–1976) arranged a large party, including an excellent live jazz band, at the Bach Dynamite and Dancing Society on the ocean at Half Moon Bay.

"Boat party, 1979." Fred Li, on the extreme right, and myself, being taught the native dance on the boat where Bill's 65th birthday party was being celebrated.

Formation of Three Rings

The critical test of an initiator is in a trienic substrate designed to
form three new rings. Our experience, as stated previously, suggested that unless an initiator gives excellent yields in bicyclizations, it will not be satisfactory for tricyclizations. Thus acetal and
certain allylic alcohol-type initiators pass this test moderately
well, as illustrated by the examples shown in Schemes 28[97] and
29.[98,99] Our lab was the scene of a great deal of excitement at the
time. Marty Semmelhack was performing the exploratory experiments on the synthetic route shown in Scheme 29 and realized the
first steroid synthesis via biomimetic polyene cyclization methodology. Even though rings A and D are opened by the ozonolysis,

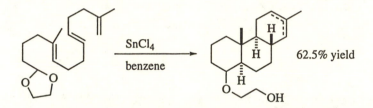

Scheme 28. Acetal-initiated tricyclization.

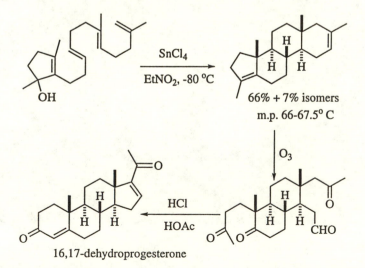

16,17-dehydroprogesterone

Scheme 29. Ditertiary allylic cation-initiated tricyclization.

the stereochemical integrity of the five new chiral centers generated in the cyclization step is still maintained in the triketo aldehyde.

Incidentally, Marty Semmelhack was the perpetrator of another form of excitement for me. He was and still is a superb tennis player and probably could have done respectably as a pro. I was no match for him in singles but on several occasions he and I played as partners in doubles matches and we have never been defeated, thanks to his skill. Once at a Gordon Conference, a couple of Canadian chemists had been beating everyone in doubles. Near the end of the week Marty and I decided to take these "experts" on and proceeded to give them a sound thrashing. The 1977 National Organic Chemistry Symposium at West Virginia University was a very special occasion for me because I received the Roger Adams Award. I had the added pleasure then of joining

Gordon Conference, 1954. (Photo courtesy of Gus Fried.)

Marty and winning a match against Don Cram and Jack Roberts. This was a rare occasion because I have almost never won any contest in which Jack was my adversary.

Schemes 30,[100,101] 31,[102–104] 32,[105,106] and 33[107] illustrate other variations of allylic alcohol initiators as well as some different

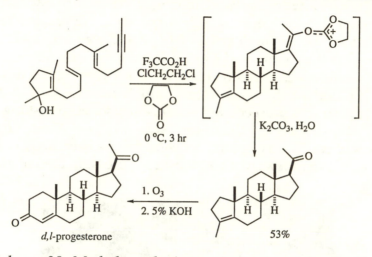

Scheme 30. Methylacetylenic terminator in a tricyclization process.

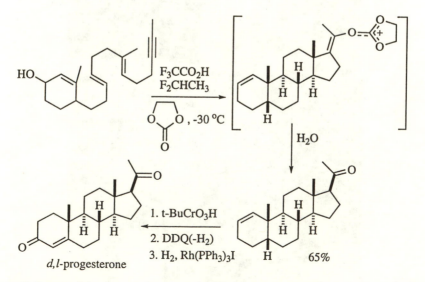

Scheme 31. Secondary–tertiary allylic cation-initiated tricyclization.

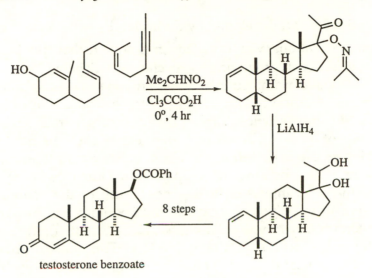

Scheme 32. Alternative application of methylacetylenic terminator with a nitroalkane serving as nucleophile.

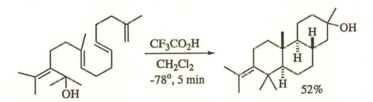

Scheme 33. Acyclic ditertiary allylic cation-initiated tri-cyclization, the Schaaf initiator.

terminators that have been used successfully to produce natural products. Other natural or biologically important products synthesized in this program include longifolene,[108a] taxodione,[108b] some 19-norsteroids,[109,110] cortisone, and spironolactone.[111]

What Makes a Good Initiator?

The epoxide group, as in 2,3-oxidosqualene, has not been a very efficient initiator of nonenzymic polycyclizations, partly because of its susceptibility to acid-catalyzed isomerization to the ketone. This problem is obviated in the case of a ditertiary epoxide initiator (*see* Scheme 34) developed by Sutherland[112] who tested it only in monocyclizations. van Tamelen[83] has shown that this initiator is quite effective in promoting a tricyclization (Scheme 35).

We have examined over 30 functional systems as potential generators of cationic initiators of polycyclizations, but only the acetal and certain allylic systems proved to be generally effective. The limits of these allylic systems were established the hard way, involving an extensive but unsuccessful endeavor, as follows.

In 1974 I became involved with a Dutch company, Organon, that was interested in the possibility of the commercial development of polyene cyclization methodology for the synthesis of steroids. My former collaborator, Koen Wiedhaup (presently Director of Research at Organon) and F. J. "Flip" Zeelen (then Director of Chemical Research), paid me a visit, with the result that an arrangement was made with Stanford for Organon to have the option to lease our patents, pending the outcome of further developments. Also, I became their consultant in this area. At that time, one of their profitable products was the 19-norsteroidal oral contraceptive, lynestrenol, which was quite expensive to produce. I suggested the synthetic route shown in Scheme 36, and with great

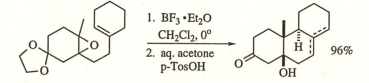

Scheme 34. *Ditertiary epoxide initiator (J. K. Sutherland et al., 1974).*

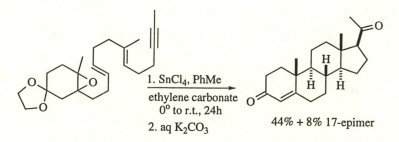

Scheme 35. *Application of ditertiary epoxide initiator to a tricyclization (E. E. van Tamelen et al., 1983).*

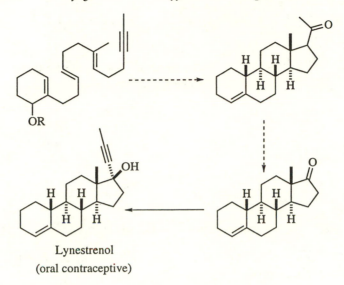

Lynestrenol
(oral contraceptive)

Scheme 36. Proposed synthesis of the oral contraceptive, lynestrenol.

enthusiasm Zeelen convinced his management to commit a number of his collaborators to this project for quite some time. Great care was given to yield optimization. They also prepared the cyclization substrate in optically active form in the hope that the enantiomeric integrity would be transmitted to the cyclization product. The surprisingly unfavorable results, which were eventually published as a joint paper[113a] are summarized in Scheme 37. The major product was bicyclic material resulting from interruption of the process after closure of one ring. A number of tetracyclic isomers were produced, the desired (normal) one being formed in only 6.5% yield. Moreover the optically active substrate underwent extensive racemization during cyclization.

Although greatly disappointing from Organon's viewpoint, these results were of theoretical importance, providing, for the first time, some insight as to the requirements of good initiators. The most important requirement is that *the initiating cation must have a critical energy, such that its bonding to the first olefinic function occurs only with the anchimeric assistance of the second (more remote) olefinic function.* If the initiator is too "hot" as with the di-secondary allylic system in the Organon example, a grossly disordered process occurs and mainly affords products (i.e., bicyclic

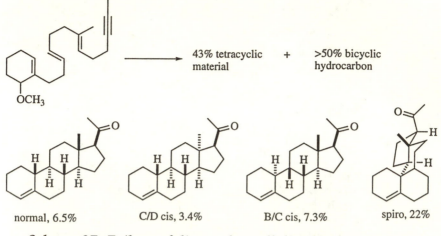

normal, 6.5%   C/D cis, 3.4%   B/C cis, 7.3%   spiro, 22%

Scheme 37. Failure of disecondary allylic cation to give selective tricyclization.

Koen Wiedhaup (Head postdoc, 1967–1968), who performed our first tetraene cyclization that proceeded in a yield of 20% which was a world record for many years. Koen and I (and families) have kept in touch ever since he left Stanford to take a position at Organon International where he is now Managing Director. Of course, my collaboration and stimulating association with the Organon people like Flip Zeelen and Marinus Groen (see pp 120–121) never would have come about if it had not been for Koen Wiedhaup. This photo was taken at Bill's home, 1987. (Photo courtesy of K. Wiedhaup.)

Koen Wiedhaup, during his days at Stanford. Noteworthy are the smile on Koen's face, the lack of safety glasses on the student, two sets of glasses, the telephone books and balance on an otherwise clean desk, and the barely discernable name tag on the front edge of the desk.

and the spirotetracycle) derived from closure of one ring (Scheme 37). In other words, the activation energy for reaction of such a destabilized cationic site with the first olefinic bond is low relative to that for the conformational reorganization of the remaining polyene chain that is necessary in the desired assisted ring closure, so that the product distribution is related to the ground-state conformations of the polyene side chain of the substrate (for a full discussion of the Curtin–Hammett principle and Curtin–Hammett kinetics, *see* the recent review[113b]). The initiating cation involved in the Organon case is evidently just on the verge of being satisfactory, because the secondary-tertiary cation involved in the examples shown in Schemes 31 and 32 is a good initiator.

Terminator Function

Although not as critical as the initiator, the terminator function
can have an important effect on the outcome of polyene cyclizations. The widely used methylacetylenic terminator, which was
developed by Mike Gravestock (postdoc. 1969–1971), has the advantage of giving the steroidal 5-membered D-ring (as in Scheme
30) with the acetyl group at C-17 (steroid numbering). Incidentally, Mike, besides being a very gifted and productive chemist
(co-author of eight papers from his two years at Stanford), is also
a talented artist. (A 4 × 5 ft. painting of his entitled "Fragmen-
tation" hangs in my office.) With the methylacetylenic terminator
the cyclizations are not totally stereo- and regioselective, yielding
some *cis*-fused as well as 6-membered ring D products as shown
in the detailed study made by Terry Lyle (Scheme 38).[114] Olefinic

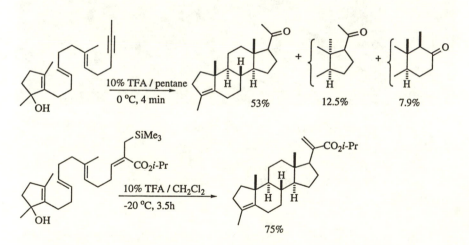

*Scheme 38. Comparison of methylacetylenic with an ole-
finic terminator.*

terminators are better in this respect as illustrated by a recent example involving the carboalkoxyallylsilane function (Scheme
38).[115] This particular terminator is useful in developing the C-17
side chain found in certain adrenal hormones (Scheme 39). Some
other acetylenic and olefinic terminators that have been examined, with the olefinic terminators always eliciting higher stereo-

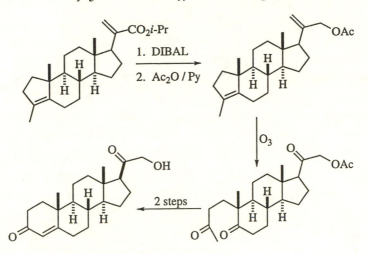

Scheme 39. Synthesis of cortexone.

selectivity, include the following with the indicated product in parentheses:

$-C\equiv CPh$ ($\rightarrow$ $-COPh$ at C-17);[116,117] $-C\equiv CCH_2SiR_3$ ($\rightarrow$ $=C=CH_2$ at C-17);[118]

$-C\equiv CSiMe_3$ ($\rightarrow$ D-homo-17-one);[119] $-CH=CHCH_2SiR_3$ ($\rightarrow$ $-CH=CH_2$ at C-17);[120]

$-CH=CFCH_3$ ($\rightarrow$ $-COCH_3$ at C-17);[114,121] $-CH=CHAr$ ($\rightarrow$ $-\overset{+}{C}HAr$ $\rightarrow$ etc.)[122]

This styryl type also includes some terminators in which the aromatic nucleus contains an ortho substituent carrying a nucleophile that intramolecularly captures the benzylic cation formed on cyclization.[123]

Formation of Four Rings

For more than 20 years efforts directed toward the nonenzymic cyclization of polyenic substrates to produce four (or more) new rings in one high yielding step have met with little success. Most attempts in this direction have given incompletely cyclized products; for example, the Lewis acid catalyzed cyclization of 2,3-oxidosqualene afforded tricyclic products with a five-membered C-ring as the only isolable polycycles (Scheme 40).[124] There have been only three partially successful experiments, as follows. The presence of pregnane-3-ol-20-one was detected in the cyclization shown in Scheme 41,[125] the low conversion probably being at least in part due to the fact that the primary epoxy function is an unfavorable (too "hot") initiator. With the acetal initiator, modest yields were obtained at best as shown in Schemes 42[126] and 43.[127] These rather unfavorable results have served to reinforce the widely accepted conviction that the objective of realizing high conversion is a hopeless one without the help of the enzyme to overcome the unfavorable entropy of activation. After a few abortive attempts to depart from performing cyclizations in solu-

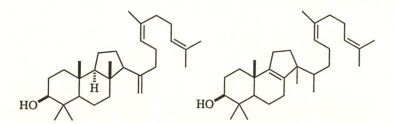

Scheme 40. Polycyclic products isolated from cyclization of 2,3-oxidosqualene with SnCl$_4$ in benzene at 10 °C for 5 min.

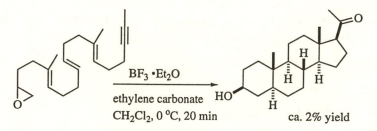

Scheme 41. Detection of natural product resulting from tetracyclization of an epoxy trienyne.

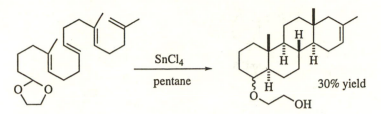

Scheme 42. Tetracyclization of a tetraenic acetal.

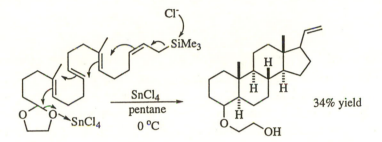

Scheme 43. Tetracyclization of a tetraenic acetal with the allylsilane terminator.

tion, such as the use of solid supports, I was inclined to consider the objective as a lost cause, until about 10 years ago (*see* subsequent section).

Thus, I was confronted with the problem of deciding whether or not to apply for competitive renewal of my NIH grant, which had already afforded 25 years of support to my polyene cyclization studies. After a great deal of deliberation about possible factors that could help to control and enhance multi-ring cyclizations, I proposed to examine substrates that would have appropriately placed auxiliaries that would stabilize cationic sites at those positions where such charges would develop in the transition state according to the Stork–Eschenmoser principle. I hoped that the effect of these auxiliaries would be to stabilize the transition state of the cyclization, thus lowering the activation energy of the desired process.

To test the concept in principle, it was decided to modify the substrate of Scheme 42 by introducing an isobutenyl group at *pro*-C-8. Thus, insofar as the bicyclic cation **56a** (Scheme 44) contributes to the cyclization transition state, the isobutenyl auxiliary

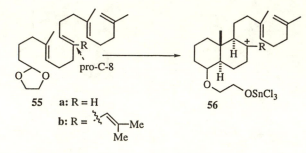

55   **a:** R = H

       b: R = $\overset{\xi}{\diagup}$ —Me
              Me

Scheme 44. Effect of a cation-stabilizing auxiliary at pro-
C-8.

in **56b** would effect allylic stabilization. Another way of viewing this system is that if the cyclization were a stepwise process, then cation **56b** would serve as a good initiator of a second bicyclization. In the event, cyclization of the auxiliary-containing substrate afforded a dramatic increase in yield, totaling 77%, of tetracyclic products, all belonging to the "natural" all-trans stereochemical series (Scheme 45).[128] This new concept, which elicited a favorable response from NIH, now shows real promise (*see* subsequent discussion). A new methodology is suggested for the stereospecific

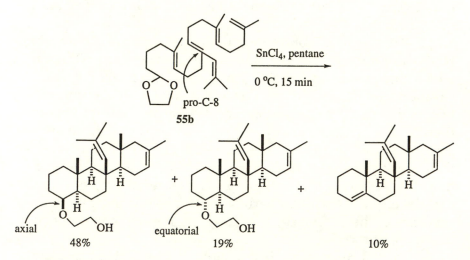

Scheme 45. Enhancement of tetracyclization by cation-stabilizing auxiliary at pro-*C-8: total yield 77% as compared with 30% for case without auxiliary.*

synthesis of polycyclic substances having four or more rings in a single step. Plans for further study include use of other auxiliaries, perhaps heteroatom types (e.g., –SR) that can be easily removed. Also it may be possible to oxidize the isobutenyl group to the angular aldehyde found in natural products like strophanthidin and aldosterone. Catalytic or photochemical decarbonylation, known to proceed with retention of configuration, could also give useful products in the case of *pro*-C-8 auxiliaries (Scheme 44). The possible biological significance of cation-stabilizing effects is considered subsequently.

Corticoid Synthesis via Polyene Cyclization Methodology

The cyclization of a substrate having a hydroxyl group at *pro*-C-11 proceeds very slowly, so that competing reactions involving destruction of the sensitive initiating function decrease the yield of the desired product (Scheme 46).[129] However, the tetracyclic material is exclusively that diastereomer with an 11-α (equatorial) substituent; hence, by employing the substrate with the indicated absolute stereochemistry, optically active cyclization product was obtained.[130] The racemic α-hydroxyalkyne intermediate shown in Scheme 46 was oxidized to the ketone (Jones oxidation), then reduced by Mosher's $LiAlH_4$–Darvon alcohol complex giving the correct (R) configuration at *pro*-C-11, in enantiomeric ratio of 92:8.[131]

Those individuals who were responsible for the pioneering experiments on the transformation shown in Scheme 46 deserve special mention. Brian Metcalf (postdoc. 1971–1972) was principally responsible for designing and reducing to practice the scheme for preparation of the racemic cyclization substrate. Ray Brinkmeyer (postdoc. 1975–1977) developed the asymmetric reduction of the acetylenic ketone to the optically active propargylic alcohol, and Sina Escher (postdoc. 1972–1974) was the first to obtain a successful cyclization (in the racemic series). The yield of the cyclization was gradually improved by others, and Terry Lyle[121] (postdoc. 1979–1980) obtained the best yield (43%) so far by performing the reaction at high dilution over a 24-h period.

On the personal side, after Brian and his artist wife, Heather Metcalf moved to Strasbourg, France, we kept in touch, seeing them several times in Europe and on one occasion touring the wine country of France together for several days. Since they returned to the States (Brian is now Senior Vice President of Chemical Research, SmithKline Beecham) we see them more fre-

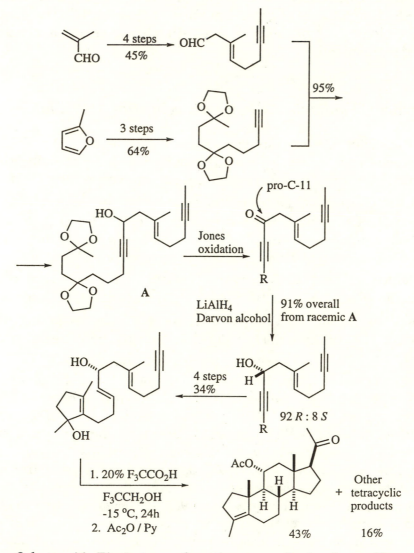

Scheme 46. First stage of cortisone synthesis: the cyclization step.

quently. Brian's career as a scientist has been spectacular. He is one of a very few administrators who at the same time is a truly outstanding creative leader in his field. When Ray Brinkmeyer was at Stanford he and I frequently played singles tennis, and we were fairly evenly matched. Occasionally Kathy and Barbara joined us in doubles.

Australian postdoc Brian Metcalf and his artist wife, Heather, who immediately fell in love with the U.S.A. and are now citizens (Stanford, ca. 1972).

The conversion of the cyclization product of Scheme 46 into cortisone acetate is shown in Scheme 47.[132] The yields of some of the steps in Scheme 47 have not been optimized; therefore, I am not yet prepared to publish the results in one of the journals. The yields for three of the conversions, marked with an asterisk, are surely subject to improvement, because three of these were only performed once on a very small exploratory scale. The yields indicated in parentheses are the values that may reasonably be expected on the basis of very closely related experiments. Unfortunately it is very difficult to persuade a new collaborator to undertake this kind of a study; indeed it would make more sense for an

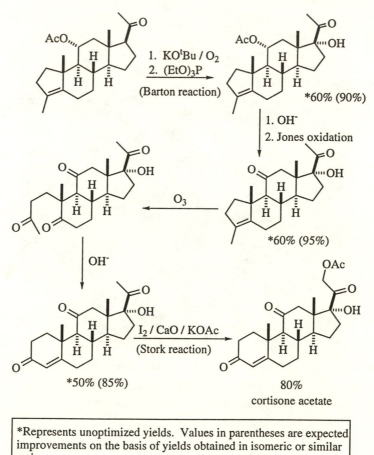

1. KO^{t}Bu / O$_2$
2. (EtO)$_3$P
(Barton reaction)

*60% (90%)

1. OH$^-$
2. Jones oxidation

O$_3$

*60% (95%)

OH$^-$

I$_2$ / CaO / KOAc
(Stork reaction)

*50% (85%)

80%
cortisone acetate

*Represents unoptimized yields. Values in parentheses are expected improvements on the basis of yields obtained in isomeric or similar series.

Scheme 47. Second stage of cortisone synthesis.

interested industrial organization to turn it over to their process development laboratories. On the basis of these reasonably anticipated improvements for four steps, the overall yield of our total synthesis of cortisone acetate would be 9% for 18 steps. Considering a cost analysis that was made when our corticoid synthesis gave 3.8% yield for 22 steps, it would appear that our present synthesis is likely to be commercially feasible. The classical Woodward synthesis (49 steps, 0.000005% yield)[36] as modified by Monsanto[133] (31 steps, 0.07%) almost went into commercial production in the mid-1950s.

Application of Cation Stabilizing Auxiliaries to Corticoid Synthesis: Discovery of Dramatic Rate Enhancement in a First-Order Assistance

The major aim of this study was to see if the cyclization of a substrate having the rate-retarding OH at *pro*-C-11 would be enhanced by a cation-stabilizing auxiliary at *pro*-C-8. At the same time we proposed to use the carboalkoxyallyl silane terminator (Scheme 38), which promised to give a product with a C-17 side chain that could be readily converted into the cortical array (Scheme 39). As shown in Scheme 48, the substrate where R = H undergoes cyclization slowly, at about the same rate as the case

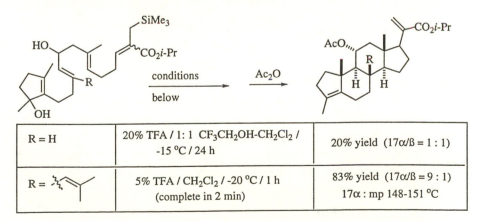

R = H	20% TFA / 1: 1 CF$_3$CH$_2$OH-CH$_2$Cl$_2$ / -15 °C / 24 h	20% yield (17α/ß = 1 : 1)
R =	5% TFA / CH$_2$Cl$_2$ / -20 °C / 1 h (complete in 2 min)	83% yield (17α/ß = 9 : 1) 17α : mp 148-151 °C

Scheme 48. First-order rate acceleration by a cation-stabilizing auxiliary: a fourfold yield enhancement.

with the methylacetylenic group (Scheme 46), and the yield is even lower, perhaps because of competing destruction of the terminator by protodesilylation.

With the isobutenyl auxiliary (Scheme 48, R = –CH=CMe$_2$) the rate of cyclization is many orders of magnitude higher, and the yield (83%) is increased fourfold.[134] This unprecedented high yield for the cyclization of a *pro*-C-11 OH polyene substrate points to the potential of applying the concept to the syntheses of corticoids. To this end, removal of the auxiliary is under investigation, as is the use of alternative (e.g., heteroatom) auxiliaries. The enormous rate enhancement represents a first-order anchimeric effect due to the cation-stabilizing (C-S) auxiliary. A second-order, greater than 10-fold rate enhancement has been observed with an isobutenyl in place of the methyl group at *pro*-C-13 of the Newton–Lindell substrate shown in Scheme 48. Hopefully, the case with the *pro*-C-11 OH will undergo cyclization giving a route to aldosterone (Scheme 49).

A Proposed Mechanism for the Action of 2,3-Oxidosqualene Cyclase

The observed effect of C-S auxiliaries on polyene cyclizations may be regarded as inferential documentation for a proposed enzyme mechanism (Scheme 50).[135] The enzyme provides external negative point charge sites[136] that stabilize, by ion-pairing, the developing cationic centers in the transition state of the cyclizing substrate. For example, the plant enzyme that produces the dammarane triterpenoids could be envisaged as delivering these point charges to the *pro*-β (axial) face of the reacting substrate, allowing for equatorial closure of the rings and extrusion of the product as the axial angular methyl groups are formed.

In the case of the liver enzyme that produces protolanosterol, the critical point charges are delivered, instead, to the α-face. Thus the boat closure of the B-ring could be promoted by delivery of a point charge to the α-face at *pro*-C-8, thereby lowering the activation energy of the boat relative to the chair closure. (The difference in activation energies for the two processes must

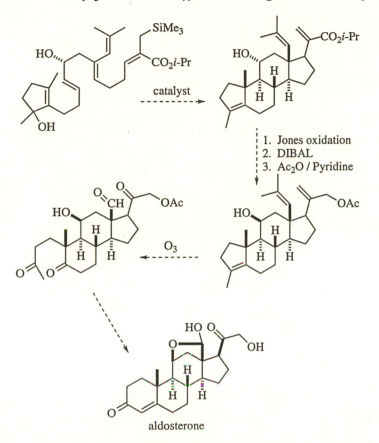

Scheme 49. Proposed application of C-S auxiliary principle.

in any case be relatively small because a significant proportion of the boat product is formed in the nonenzymic cyclization of a squalene epoxide analog.)[137] The ubiquitous but enigmatic non-Markovnikov biological closure of the C-ring may be favored by delivery of a point charge at *pro*-C-13. (Note that the *non*enzymic cyclization of epoxysqualene gives 5-membered ring C products. *See* Scheme 40.) Thus the enzyme, instead of having to impose a profound and highly restrictive conformational influence on the substrate, may provide an electronic environment that guides and sustains processes which are already inherent in it: that is, the Stork–Eschenmoser principle. An attractive feature of this model is its simplicity.

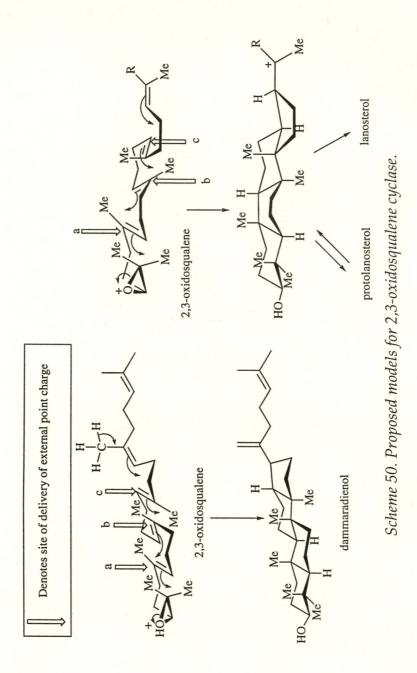

Scheme 50. Proposed models for 2,3-oxidosqualene cyclase.

The proposed enzyme models for these two cases are shown in Scheme 50. Stabilization by charge a, delivered to the β-face at *pro*-C-10 is regarded as providing a powerful first-order effect helping the epoxy function (which is not a particularly good initiator in vitro) to perform as an efficient initiator, catalyzed by a weak (possibly a carboxylic) acid. Charge b, directed to *pro*-C-8, determines whether ring B closes as the chair or boat depending on whether delivery is to the β- or α-face of the substrates, respectively. Finally, charge c, directed to *pro*-C-13, is responsible for the non-Markovnikov closure of ring C.

I assume full responsibility for perpetrating this speculative concept; however, I have had helpful discussions with a number of experts who have supplied some touches that I have adopted. My colleague John Frost called my attention to the rhodopsin model, and Barry Carpenter (Cornell) suggested the lysozyme model to me as well-documented examples of enzyme mechanisms involving negative point charges. Other friends who examined my hypotheses and encouraged me to stick my neck out include Konrad Bloch, John Brauman, Albert Eschenmoser, Koji Nakanishi (who has also published his autobiography, *A Wandering Natural Products Chemist*, within this Profiles, Pathways, and Dreams series), Jack Roberts, and Frank Westheimer. I learned from John Cornforth that as long ago as 1959 he suggested,[138] on the basis of pure intuition, the possibility of "complexation" of the enzyme with the developing positive centers of the substrate. Also Guy Ourisson recently has been thinking in terms of "nucleophiles" being delivered by the enzyme.[139]

Stereoselective Olefin Syntheses

Our extensive studies of biomimetic polyene cyclizations involved, as a major part of the total effort, the synthesis of polyenic substrates with all *E*-olefinic bonds. At the start of our cyclization studies in the early 1960s, there were very few stereoselective, high-yield methods for producing trisubstituted olefinic structures, and these are briefly reviewed in one of our 1968 articles.[140] Consequently, we became involved in developing new methodology.

At my desk.

Brady–Julia Synthesis of Trisubstituted Olefins

In 1960 Marc Julia and his collaborators showed that cyclopropyl-carbinols, on treatment with 48% HBr, are converted into homoal-lylic bromides.[141] The olefinic bonds of these products were formed stereoselectively: 90–95% *E* for the disubstituted case, but only ca. 75% *E* for the important trisubstituted case (*see* Scheme 51). My student Steve Brady (Ph.D. 1967), as part of his thesis project, undertook a study with the aim of improving the stereose-lectivity of the Julia synthesis for trisubstituted olefins. As it turned out, part of our success probably was the result of a struc-tural modification of the substrate so that a secondary rather than

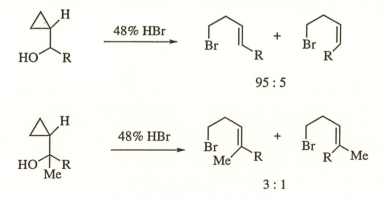

Scheme 51. Julia olefin synthesis.

a tertiary carbinol was used. The secondary carbinol structure appears to have a larger difference in energies between the two transition states (*see* Scheme 52). In addition, it was expected that avoiding the strongly acidic conditions would be beneficial. Without any help from me, Steve dug through the literature and located, in particular, a paper of Jack Roberts[142] showing that both cyclopropylcarbinyl and cyclobutyl bromides were converted into the homoallylic bromide with zinc bromide.

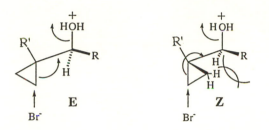

Scheme 52. Possible models for transition states of Brady–Roberts rearrangement.

Steve eventually worked out a nice procedure involving conversion of the carbinol, by treatment with PBr_3 and LiBr in collidine at 0 °C, into a mixture of the two cyclic bromides, which were then rearranged by stirring with $ZnBr_2$ in ether at 0 °C.[140] Thus trisubstituted olefins were produced with 98% *E* selectivity. We have found this method to be of great practical value for the stereoselective synthesis of substrates for polyene cyclizations. An example is the synthesis (*see* Scheme 53)[143] of the substrate **57** for the asymmetric cyclization described on page 150.

In addition to using the method for producing a number of our polyene cyclization substrates, the Brady–Julia reaction has been applied to the synthesis of some olefinic natural products (i.e., dendrolasin)[144] and the first stereoselective syntheses of the C_{18}[145] and C_{17}[146] cecropia insect juvenile hormones (JH). The synthesis of C_{18} JH is suggested in Scheme 54, the process proceeding from right to left. Except for the last step, the new C–C bonds are formed by the successive alkylations of enolate anions.

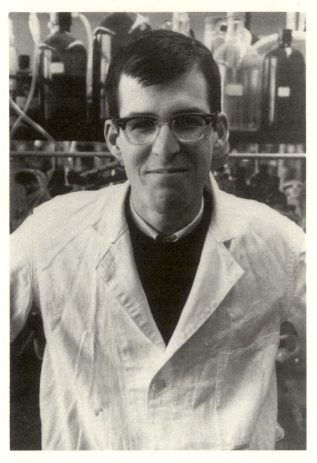

Steve Brady, in my laboratories at Stanford, 1967.

The Ortho Ester Claisen Reaction

Probably the most useful chemistry to emanate from our laboratories is the ortho ester Claisen reaction, which is illustrated by
the example in Scheme 55. A computer literature search in the *Science Citation Index* for references citing our seminal paper[147] on the
subject between 1974 and 1993 turned up 331 papers, with 24 and
22 citations in 1994 and 1995, respectively. An additional 54 citations have been found between 1970 and 1974 in the *Cumulative
Index* of *Chemical Abstracts*.

In the 1960s, our need for a practical stereoselective synthesis of *E*-trisubstituted olefinic systems was inspired by our aim

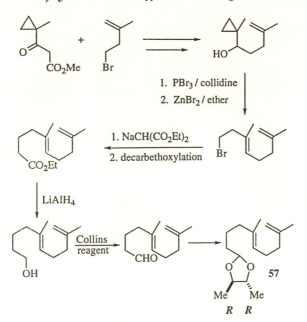

Scheme 53. Application of Brady–Julia olefin synthesis to the preparation of a cyclization substrate.

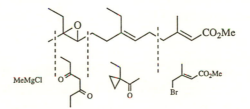

Scheme 54. Application of Brady–Julia olefin synthesis to the first stereoselective synthesis of Cecropia juvenile hormone.

of producing polyene substrates like those shown in Scheme 42. When we tried to optimize the selectivity of the classical Claisen reaction of the vinyl ether **60**, the best we could do was the indicated 86:14 *E/Z* ratio of aldehyde **61**. When I suggested that ethyl orthoacetate might be used directly with an allylic alcohol like **58** to give the product **59**, it had not occurred to me that this process would show improved selectivity. Indeed, I was completely (and

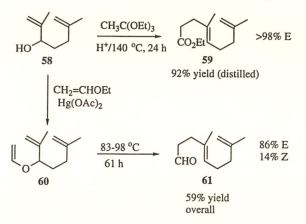

Scheme 55. Comparison of orthoacetate Claisen rearrangement with classical Claisen rearrangement.

pleasantly) surprised that the reaction proceeded so highly in favor of the E product.

An account of the discovery of the ortho ester Claisen reaction follows. My predoctoral student, William (Ted) Bartlett, who had just about completed his work for the Ph.D. in the spring of 1969, was devoting his major effort to exploratory experiments on the problem at hand. I asked Ted if he would give me an account of his recollection of the matter and on March 10, 1993, he wrote to me as follows:

> My recollection of the development of these Claisen rearrangements is that we were all teaching each other and I was applying new ideas at the bench as we developed them together, trying to discover the secret of how to increase the stereoselectivity of the rearrangement....I seem to recall telling you about the successful amino ketal rearrangement and your saying "If that rearrangement works, I wonder if the ortho ester version would also work?" Do you remember that conversation?

My answer was in the affirmative; indeed, I remember feeling that if the ortho ester failed to work we could always fall back on the more complex Eschenmoser method, which gives excellent selectivity.

Kathy Bancroft, Ted Bartlett, and Soan Cheng at the 1995 Johnson Symposium. (Photo courtesy of Eric J. Leopold)

Ted also sent me a copy of page 137 of his notebook (*see* Figure 3), which displays a record of the first ortho ester Claisen experiment. Note that it works very well even though no acid catalyst was used as in the Eschenmoser recipe. Ted also recorded some results of early experiments of other people on this page.

After the fact, my collaborators Fred Li and John Faulkner rationalized the result as being due to a considerable nonbonded interaction between the EtO and R group that develops only in the transition state leading to the Z product (Scheme 56). When it comes to publishing or lecturing about work that develops in this way there is a great temptation to dramatize the story by presenting the rationale first (as though it were part of the conception), followed by the experimental results. There is no way of apprehending this mild form of dishonesty, which I suspect is rather widely practiced.

Before moving on to another topic I would like to say a little more about Ted Bartlett, who in addition to performing the first example of the ortho ester Claisen reaction has made a number of other novel contributions to our program. Ted's Ph.D. thesis was so well written that I suggested that he try his hand at

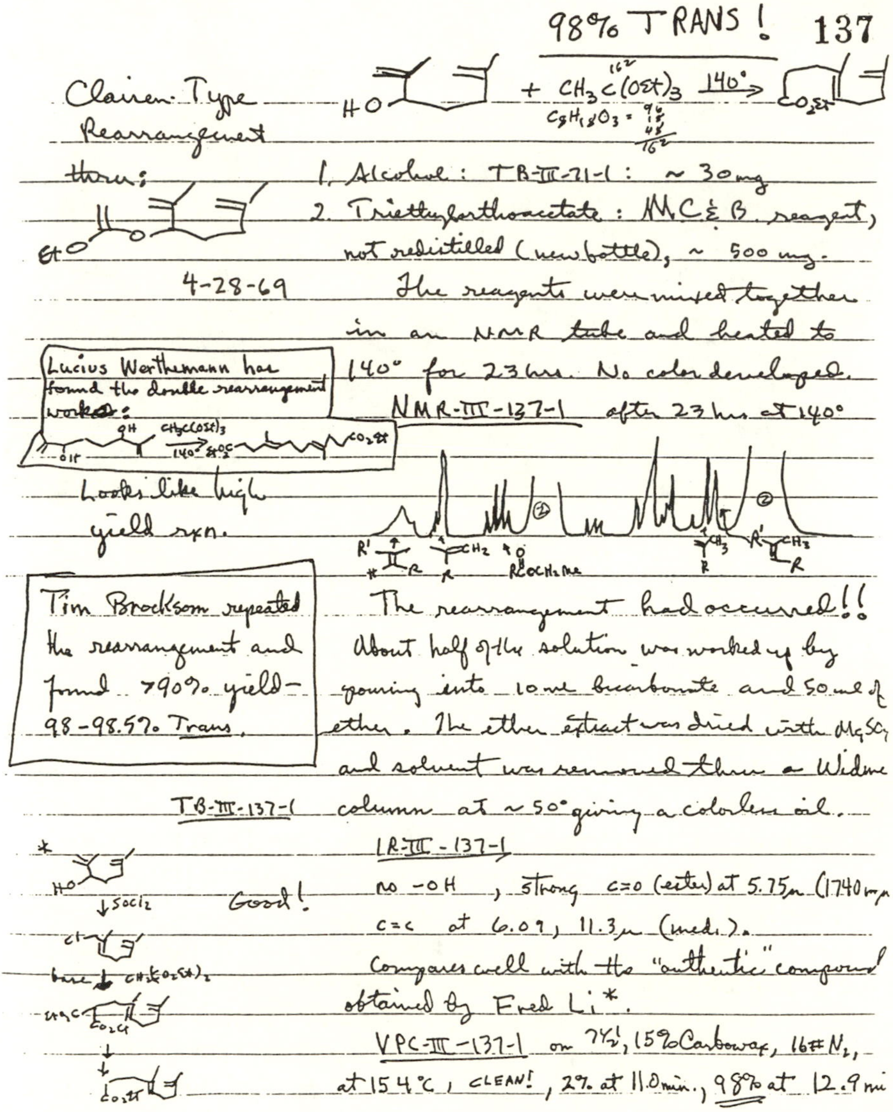

Figure 3. The April 28, 1969 laboratory notebook page of Ted Bartlett's describing the first ortho ester Claisen experiment.

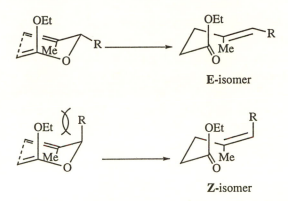

Scheme 56. Models for transition states leading to E- *and* Z-*products in orthoacetate Claisen rearrangement.*

writing it up for publication. The result was splendid—something I have rarely experienced with green Ph.D. students.

Ted always had a strong desire to teach in a liberal arts college, but at the same time he was determined to continue to be very active in research. Hence after a postdoctoral year with Gilbert Stork, and short stints at a couple of Midwestern colleges, he found what he wanted at Fort Lewis College where he could work closely with and inspire students. At the same time, he has maintained serious activity in research, which has also been to my advantage because, in addition to his own independent programs, he has continued to collaborate with us, particularly during sabbatical leaves. In this connection, he has also done some major preparation of a number of manuscripts for us. It is a great pleasure and advantage to have a "permanent" collaborator of this sort.

The ortho ester Claisen reaction has proved to be quite popular because it is very easy to perform by using a common reagent from the shelf, and it gives high yields of a clean product with very high *E* selectivity, usually >98%. In our original 1970 paper[147] we also reported the use of the methodology to produce squalene via a two-directional symmetrical process starting with succinic dialdehyde. Another phase of this paper concerned the use of 3-methoxyisoprene in the Claisen rearrangement to synthesize squalene. John Faulkner, with his student Michael Petersen, had become involved in this chemistry[148] after moving to the Scripps Institute, and we collaborated on its application to the squalene synthesis.

　　We have used the ortho ester Claisen reaction extensively for the synthesis of polyenic cyclization substrates (*see*, for example, Scheme 57)[128]. Another application was in the synthesis of some pheromones of the Queen and Monarch butterflies, a joint project with Jerry Meinwald.[149]

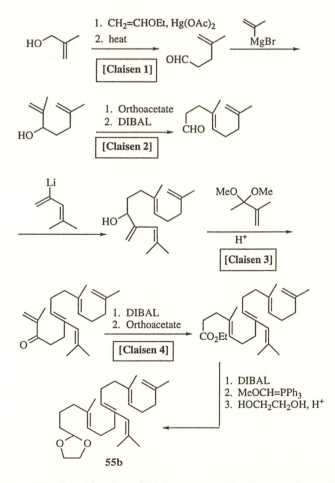

Scheme 57. Synthesis of polyene cyclization substrate via four Claisen rearrangements. DIBAL is diisobutylaluminum hydride.

Olefinic and Chloro Ketal Claisen Reactions

The olefinic ketal Claisen reaction,[150] in its simplest form, involves an intermediate like that shown in Scheme 56, except that the ethoxy group is replaced by the vinyl group, which leads to an α,β-unsaturated ketonic product (Scheme 56, the Z-isomer where EtO is replaced by a vinyl moiety). The reaction is carried out similarly to the ortho ester Claisen process except that the ketal of an α,β-unsaturated ketone is used. This reaction was discovered in 1967 by my extraordinarily talented collaborator, Fred Li, who heated the ethylene glycol ketal of mesityl oxide with geraniol and obtained an α,β-unsaturated ketonic product. The enol ether (e.g., methoxyisoprene as discussed previously) can be used as an alternative. A recent application of this methodology is found in the synthesis of the cyclization substrate of Scheme 45. This synthesis was performed by a sequence of Claisen reactions, the second and fourth being ortho acetate and the third an olefinic Claisen reaction (Scheme 57).[128] An improved synthesis of juvenile hormone, based on the olefinic ketal Claisen reaction, was reported in the seminal paper announcing the new methodology.[150]

The chloro ketal Claisen reaction was developed as a result of a suggestion made by the late John Siddall (Zoecon). It involves the use of the ketal of a tertiary α-chloroketone in the Claisen reaction with an allylic alcohol (**62 + 63 → 64,** Scheme 58). Norman Hunter (postdoc. 1970–1971) conceived of the application of this reaction to a facile synthesis of the Schaaf cyclization initiator (Scheme 33) as shown in Scheme 58. This methodology has been used in the total synthesis of the pentacyclic triterpenoid, serratenediol (Scheme 59). This initiator was employed in a cyclization to produce the D/E rings of substrate **67** as well as the A/B/C rings in the cyclization **67 → 68**. The basic plan for this synthesis was suggested by Jeff Labovitz (postdoc. 1973) who, along with Glenn Prestwich (Ph.D. 1974), reduced the idea to practice. My part in the project was so small that I encouraged them to publish the work independently.[151]

The product of the chloro ketal Claisen reaction can be readily converted to the epoxide (Scheme 60), thus providing a

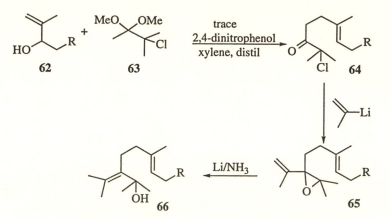

Scheme 58. Hunter's approach to Schaaf cyclization initiator.

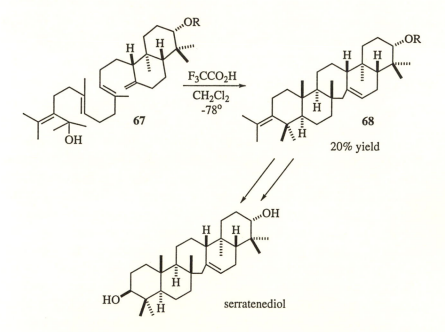

serratenediol

(9 asymmetric centers, 512 optical isomers, 256 racemates)

Scheme 59. Labovitz–Prestwich synthesis of a pentacyclic triterpenoid.

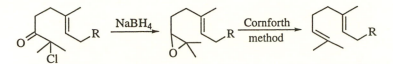

Scheme 60. Use of chloroketal Claisen product to generate epoxy residue of oxidosqualene or of juvenile hormone, as well as to form the terminus of squalene.

very efficient way of terminating the juvenile hormone synthesis.[152] By using the optically active form of the ketal of Me-COCClMeEt (compare **63**) in this synthesis, the optically active form of C_{18} JH was produced.[153] Furthermore, the epoxide can be converted, reductively, to the isopropylidene group. Thus the chloro ketal Claisen methodology has been applied to the termination of the aforementioned total synthesis of squalene, resulting in a highly efficient process.[154]

Enantio-Controlled Generation of Chiral Centers During C–C Bond Formation Mediated by Homochiral Acetals

In 1966, without any great expectations, I thought it would be interesting to examine a cyclization like that in Scheme 24 (which gives a racemic product), except that the acetal would be derived from an optically active diol, namely 2,3-butanediol, which was available in the *R,R* form as a fermentation product. Thus, the cyclization shown in Scheme 61 was performed and the products, after removal of the chiral auxiliaries (via ether cleavage), were of fairly high enantiomeric purity. However, the absolute configurations at the ring fusion were opposite in the axial (formula **69**) and equatorial (formula **70**) diastereomers, as shown by conversion to the ketones, which proved to be enantiomeric.[143] The ratio of **69** to **70** was only about 5:2. Unfortunately, with a cyclization solvent like nitromethane, which gives high diastereoselectivity for the axial epimer, the enantioselectivity was poor; therefore, we did not regard this as useful methodology.

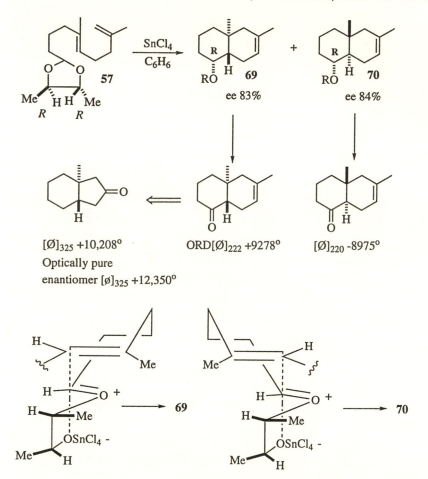

Scheme 61. Enantioselective cyclization mediated by homochiral acetal template and proposed transition states.

Nothing further was done with the problem until 1981,
when a new study was initiated that culminated in the case
shown in Scheme 62. The conditions for high diastereo- as well as
enantioselectivity were developed by John Elliott (postdoc. 1981–
1983), who also was responsible for proposing as well as reducing
to practice the use of the acetal derived from 2,4-pentanediol. This
methodology has the enormous advantage of giving products (**72**
and **73**) that allow for very facile removal of the chiral auxiliary by
oxidation of the product alcohol to the ketone followed by either
base- or acid-catalyzed β-elimination (**74** → **75** → **76**, Scheme 63).

In contrast the removal of the auxiliary from **69** and **70**, R = OCHMeCHMeOH, is capricious and the yields are generally poor to fair. The case in Scheme 62 proceeded in high yield and diastereo- and enantioselectivity. The major product was easily converted into the Inhoffen–Lythgoe diol, 96% ee, an important in-

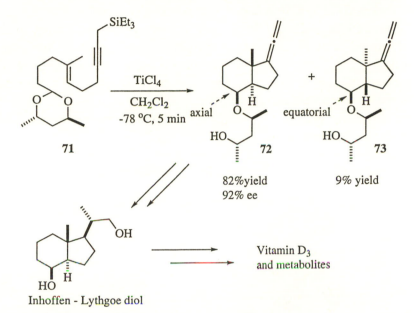

Scheme 62. Application of asymmetric cyclization to synthesis of intermediate for vitamin D₃ types.

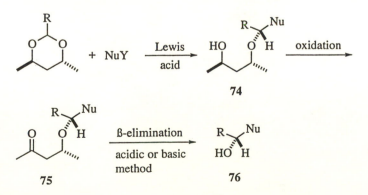

Scheme 63. Generalized intermolecular reaction of homochiral acetals with nucleophiles.

termediate in the synthesis of vitamin D_3 metabolites such as calcitriol.[155]

In 1982, Paul Bartlett sent me a copy of a manuscript he was preparing for his chapter in Morrison's book,[86] and we were intrigued with a mechanism he was proposing for the cyclization of Scheme 61. He pointed out that while the axial and equatorial products **69** (R = H) and **70** (R = H) are of opposite chirality at the bridgehead centers, these two epimers have the *same* configuration (*R*) at the carbinol center. The chiral auxiliary therefore exerts a strong influence only on the direction of attack by the nucleophile (olefinic bond) on the cationic acetal Lewis acid complex, favoring the *R* chirality at the center irrespective of the stereochemical sequelae.

This analysis made me feel that similar stereoselectivity might be realized in the *inter*molecular counterpart. In discussing

W. S. Johnson and John Elliott, making enantioselective plans.

this idea with John Elliott, a plan was developed to test the concept by using his type of acetal derived from 2,4-pentanediol with various nucleophiles as generalized in Scheme 63 for the *R,R*-acetal. The favoring of the formation of **74** rather than the diastereoisomeric ether may be rationalized on the basis of an S_N2-like transition state **77** (Scheme 64), which is stabilized by a lengthening of the 2,3-bond (process a) of the ground state **78** with consequential relief of the relatively large (in the dioxane system) 2,4-diaxial H/Me interaction. No such interaction is relieved in the alternative process b involving 1,2-bond lengthening in the transition state **79**, which is therefore less favored.

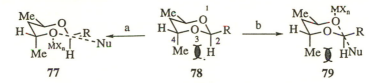

77   **78**   **79**

Scheme 64. Possible transition states for coupling reaction of Scheme 63.

John performed a number of preliminary experiments that looked very promising, then almost everyone in our group became involved in some phase of this program. I persuaded Paul to join us in the seminal paper[156] disclosing his mechanistic concepts, along with John's very successful early experiments on the coupling with allyltrimethylsilane to give (ultimately) optically active homoallylic alcohols. Later, further refinement of the latter study resulted in a process that gave these alcohols in very high yield and ee as shown in Scheme 65.[157]

Scheme 66 shows some of the nucleophiles studied with the final products obtained. Note that 2,4-pentanediol is available in the *S,S* as well as the *R,R* form (Aldrich); hence, the enantiomeric forms of all products of Scheme 66 are available. Use of methallyltrimethylsilane (**80**, Scheme 66) with the acetal where R = 3-butenyl gives an alcohol (ee >98%) that was converted into the natural product (–)-*dihydromyoporone*.[157] The methallylation was also used in a synthesis of the key intermediate for the preparation of the vitamin D_3 metabolite, calcitriol lactone.[158]

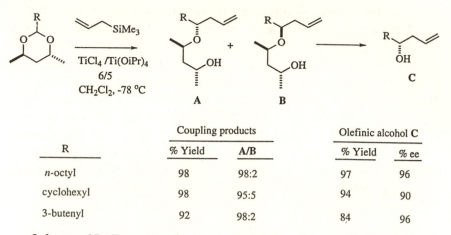

R	Coupling products		Olefinic alcohol C	
	% Yield	A/B	% Yield	% ee
n-octyl	98	98:2	97	96
cyclohexyl	98	95:5	94	90
3-butenyl	92	98:2	84	96

Scheme 65. Enantioselective synthesis of homoallylic alcohols.

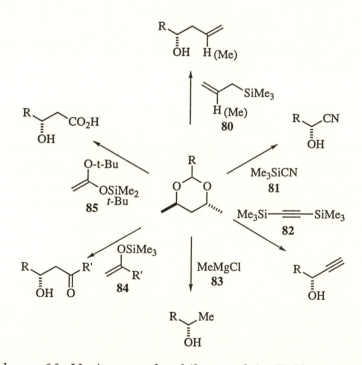

Scheme 66. Various nucleophiles used in TiCl$_4$-promoted couplings of chiral acetals after removal of the chiral auxiliary.

Trimethylsilyl cyanide (**81**, Scheme 66) was suggested by John Elliott, who is the principal author of the publication.[159] This study resulted in an expeditious asymmetric synthesis of cyano-hydrins. An example of the importance of this discovery is that the *S*-cyanohydrin with R = *m*-phenoxyphenyl can be made by this methodology.[160] This *S*-cyanohydrin is the alcoholic compo-nent of most of the important pyrethroid insecticides.

The use of bistrimethylsilylacetylene (**82**, Scheme 66) leads to propargylic alcohols, and various compounds were made that are optically active intermediates in the synthesis of such prod-ucts as the pheromone of the dried bean beetle, certain of the prostaglandins, and α-tocopherol.[161] The use of organometallics, exemplified by the case shown at **83**, Scheme 66[162] has led to the synthesis of a number of optically active carbinols including a novel pyrethroid alcoholic component, R = *m*-phenoxyphenyl, that gave a pyrethroid having strong activity against the common housefly.

The aldol coupling with enol silanes (**84**, Scheme 66)[163] proceeds in high yields and ee. It is not possible to remove the chiral auxiliary, as implied, because the β-elimination step also destroys the product by further β-elimination. This problem was overcome by using the acetals from 1,3-butanediol, which are available in both *R* and *S* forms. Coupling reactions occur regiose-lectively to open the acetal so as to give the primary alcohol.[160,164] The aldehyde obtained on pyridinium chlorochromate (PCC) oxi-dation undergoes facile β-elimination under mild conditions (dibenzylammonium trifluoroacetate) giving the aldol product in high yield and ee. This chemistry has been applied to the prepa-ration of key intermediates in Bartlett's nonactic acid synthesis. The related use of ketene acetals (**85**, Scheme 66) has been applied in the key step of an efficient asymmetric synthesis of (+)-α-lipoic acid.[165]

It should be emphasized that John Elliott has played a major role in this program intellectually as well as at the bench. Since leaving Stanford (now at SmithKline Beecham), he has kept up with the field and has continued to provide valuable input to our program in this area. Extension and development of the methodology as well as its application to new situations are under investigation. A number of scientists in various parts of the world are now also involved in such studies, and the names of some of these individuals may be found in our 14 papers.

Also Paul Bartlett's input was a very important factor in this work as well as in the biomimetic polyene cyclization studies. Ever since Paul was a student at Stanford, I have enjoyed a close relationship with him.

Fluorine Atom as C-S Auxiliary in Biomimetic Cyclizations

Our most recent studies have been concerned with this topic. Although the isobutenyl group was an effective C-S auxiliary (*see* Schemes 44 and 45), the cyclization products could not be converted into synthetically useful species because the isobutenyl group in the cyclized product could not be cleaved.[134] The fluorine atom on the other hand has proved not only to be an effective C-S auxiliary, but it can readily be eliminated from the product (giving an olefinic bond) or it can be reductively displaced by hydrogen with retention of configuration. Both of these features are useful.

The total asymmetric synthesis of a steroid, which was performed by my co-workers Balan Chenera and Vernon Fletcher, is depicted in Scheme 67.[166] The cyclization substrate **86** with the *S,S* acetal initiator and a fluorine atom at *pro*-C-8, on treatment with $SnCl_4$ in dichloromethane at –90 °C, gave a 69% yield of tetracyclic material, the major component of which was **87**. (The remaining products were mainly those arising from dehydrofluorination catalyzed by the Lewis acid.) Treatment of **87** with the Ohsawa–Oishi reagent (Na/K alloy and crown ether in toluene at 23 °C) resulted in replacement of the fluorine atom by hydrogen, with retention of configuration, to give **88**, which was readily degraded to the known steroid **89** of high optical purity. This material was compared directly with the naturally derived product.

The cation-stabilizing effect of the fluorine atom has not only proved effective in enhancing the yields of polycyclizations but it has also been very useful in controlling the regioselectivity

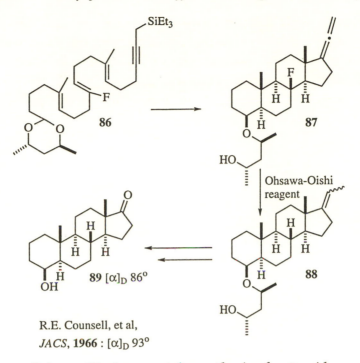

R.E. Counsell, et al,
JACS, **1966** : [α]$_D$ 93°

Scheme 67. Asymmetric synthesis of a steroid.

of the process. As indicated by Scheme 68 the enzymatic cycliza-tion of oxidosqualene almost always proceeds so as to form **91**, a six-membered ring C, in opposition to **92**, the pathway predicted by Markovnikov's rule. The nonenzymatic (acid-catalyzed) cycli-zation of the oxide on the other hand generally proceeds in the manner predicted by theory giving a five-membered C ring. Thus, Albert Eschenmoser (with tongue in cheek) would call the enzy-matic process "nonchemomimetic". The regiochemical problem can be beautifully controlled by the use of an appropriately placed fluorine atom in the cyclization substrate as shown by the basic studies of my co-worker Robert (Bob) Buchanan.[167] This methodology is impressively illustrated by the total synthesis of β-amyrin, which was reduced to practice by my co-workers Mark Plummer and S. Pulla Reddy (Schemes 69 and 70).[168]

The synthesis of the cyclization substrate **93** was fairly straightforward except for the formation of the tetrasubstituted fluoroolefinic bond in the required configuration. After a number

of attempts with various forms of the Claisen and Wittig rearrangements we discovered that application of the Trost π-allyl-Pd
reaction with carbon nucleophiles as depicted in Scheme 69 gave
very gratifying results.[169] Thus this strategy, which has elements
of convergency, introduced an incipient cyclopropyl carbinol

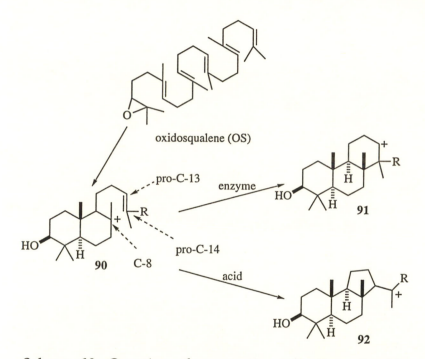

Scheme 68. Question of enzymatic cyclization of oxido-squalene being chemomimetic.

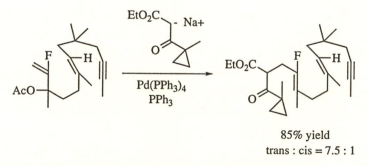

85% yield
trans : cis = 7.5 : 1

Scheme 69. Application of Trost reaction.

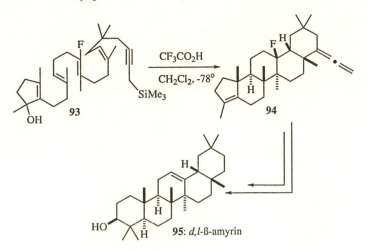

Scheme 70. Total synthesis of d,l-β-*amyrin.*

function required for subsequent formation of the 3-trans olefinic bond via the Brady–Julia rearrangement (*see*, e.g., Scheme 53). The cyclization of **93**, Scheme 70, with trifluoroacetic acid in dichloromethane at −70 °C gave the surprisingly high yield of 64–70% of the pentacycle **94**. Thus, the fluorine atom not only enhanced the cyclization, but it controlled the regiochemistry so as to give exclusively the 6-membered ring C. The transformation required for conversion of **94** into racemic β-amyrin is fairly obvious. The fluorine atom in cyclization products like **94** is relatively stable in the presence of protic acids, but it is readily eliminated on mild treatment with Lewis acids. $SnCl_4$ was used for the regiospecific introduction of the olefinic bond at C-12, at an appropriate point, in the conversion of **94** to **95**.

Some Concluding Remarks About Our Research

The reviewed research did not evolve from a master plan; indeed it was largely a matter of following one's nose and trying to look at things that related to areas that were regarded as important issues of the time. Almost any organic reaction fascinated me, particularly if it involved the formation of new carbon–carbon bonds and hence could be applied to synthesis. When I began independent research, steroid total synthesis, then in its infancy, was a challenging area of importance. At the beginning I had the desire to do something significant in this field but had no really good ideas. Instead I turned my attention to the Stobbe condensation, partly out of interest per se and partly because it afforded the possibility of introducing a propionic acid side chain at the site of a keto carbonyl, a transformation required in the Bachmann equilenin synthesis. It was a matter of pure luck that our attempts to apply the Stobbe condensation to a Bachmann-like intermediate resulted in a stereoselective synthesis of equilenin (Scheme 4).

When it was found that the hydrochrysene derivative 34 (Scheme 10) could be readily produced in quantity, it was rather obvious to try to use this intermediate for elaborating steroidal structures. Once the stereoselective methods for introducing the bridgehead chiral centers were developed, it was relatively easy to plan a major program for the synthesis of a number of steroids. Again, a large proportion of this program was performed without the aid of NMR spectroscopy.

Our largest 35-year effort concerning biomimetic polyene cyclization began with some probing experiments, without any real feel for where we were going. When the sulfonate ester solvolysis studies began to look quite hopeless, we would have abandoned the whole project had we not luckily tried acetal and allylic alcohol functions as cyclization initiators without any feeling of assurance that they might lead to success. Thus, the stage was set for examining a variety of initiators as well as terminator functions, learning how to make polyene substrates stereoselectively, and then applying these findings to the total synthesis of polycyclic compounds, particularly the steroids.

Others in addition to van Tamelen, who made major contributions to the field of biomimetic polyene cyclizations include F. J. Zeelen,[87,170] W. N. Speckamp,[171] and H. M. Buck[172] along with their respective collaborators. In particular the Zeelen group improved and greatly extended the scope of the Bartlett estrone synthesis (Scheme 26). Buck et al. produced interesting sulfur heterocycles by using thiophene residues in place of the aryl group in this type of synthesis, and the Speckamp group discovered a new and efficient initiator function, the acyliminium cation, that leads to nitrogen heterocycles including aza steroids. On February 24, 1977, these Dutch scientists participated in a symposium on polyene cyclizations at Organon arranged by Flip Zeelen to celebrate my 64th birthday, a delightful occasion for me. All of my Dutch (academic) grandchildren attended, and I was presented with a pair of appropriately engraved wooden shoes that have since been on exhibit in my office.

I never would have guessed that the polyene cyclization program would still be one of our major efforts after 35 years. My father used to say that there are two ways to write a book, one is "to a plan", the other is "from a plan". He always opted for the former strategy, letting the plan evolve as he wrote. For the most part, our research likewise has been performed to a plan: We sort of followed our noses addressing new problems that presented themselves after giving a great deal of thought to immediate priorities. This is not to say that we did not do a great deal of serious planning of specific projects even though these plans were often thwarted (see the "fallibility phenomenon" discussed later). However, there was no great master plan, although it is possible to construct one after the fact. Until very recently we had no idea that we would be involved in trying to synthesize partially cyclized substances as transition-state analog enzyme inhibitors. These substances contain atoms carrying a positive charge located at sites where positive charges are presumed to develop in the transition state of the cyclization of oxidosqualene. Thus we find ourselves involved in a collaborative effort with an enzymologist, Pierre Benveniste (Strasbourg), in an effort to test our enzyme mechanism.

The polyene cyclization studies also prompted us to address the issue of new methodologies for stereoselective olefin syntheses, which continues to be an important area of investigation. Applications to synthesis of insect hormones and pheromones were obvious, timely spinoffs. Another spinoff was derived from the observation that initiation of a polyene cyclization by a homochiral acetal resulted in an asymmetric synthesis of a polycycle; the resulting study of reaction of various nucleophiles with chiral acetals has developed into a major field of its own.

Throughout the survey of our research presented in the previous sections, I have mentioned certain individuals who have had an especially strong influence on my own studies. I was pleased to have the opportunity recently to pay tribute to these people in the following way. When the Stanford Chemistry Department decided, as the result of a proposal made by Carl Djerassi, to establish the "William S. Johnson Symposium in Organic Chemistry", I was asked to deliver a talk as well as to select the other speakers for the first event in 1986. This proved to be a simple task for I chose those aforementioned special individuals: namely, Derek Barton, Konrad Bloch, Albert Eschenmoser, Jack Roberts, and Gilbert Stork, all of whom honored me by participating. Had Werner Bachmann been alive, he also would have been included. A large number of my former as well as present collaborators attended this affair, which was a deeply moving occasion for me.

In 1986, Carl Djerassi and Paul Wender raised the money for initiation of what they called "The William S. Johnson Symposium in Organic Chemistry." Since then, Carl, Paul and Barry Trost have been responsible for continuing the symposium [now into the 11th year]. They asked me to select the speakers for the first symposium which was an easy task for I simply chose those whose work had made the greatest influence on my own. These people also happen to be excellent speakers. The organizing committee also insisted that I be one of the first speakers which gave me the opportunity to thank everyone who was involved in the whole affair. The speakers for the first symposium are shown in the photograph from left to right as follows: Albert Eschenmoser, myself, Jack Roberts, Gilbert Stork, Derek Barton, Carl Djerassi (Program Director), and Konrad Bloch. (Photo courtesy of Eric J. Leopold.)

W. S. Johnson at the first Johnson Symposium, Stanford, 1986.
(Photo courtesty of Eric J. Leopold.)

Industrial Associations

My contacts with industrial chemistry began while I was still in school and have continued to be an important part of my professional as well as personal life ever since. These associations brought me into contact with a number of very talented scientists from whom I learned much chemistry, some of which had an impact on my own research program. My industrial contacts also proved to be important in helping my own collaborators to obtain appropriate permanent positions. The strongest associations have been with the companies mentioned below.

My employment at the Eastman Kodak Company during the summers of 1936–1939 already was described. It was at Kodak that I became interested in the Combe's quinoline synthesis (Scheme 2), which was studied by my first Ph.D. student.

Consulting

In 1945, Merle Suter of the Winthrop Chemical Company invited me to become a consultant, and I continued in this capacity for 24 years to see the organization grow into the Sterling-Winthrop Research Institute, a major pharmaceutical research facility. Merle was a fine administrator who really cared about people and I enjoyed a warm friendship with him and his wife Verle. I had many friends at Sterling, among whom were (eventually) nine of my former students and postdoctoral associates, three of whom have reached managerial positions. My three to four annual visits were important social as well as professional occasions, which afforded me the opportunity also to see my brother, Tom (and his wife Miriam) who obtained his Ph.D. with Al Wilds in 1946 before accepting a position at Sterling.

Professionally my interactions with Sterling were very good for me and, at least to some extent, for them also. Sterling supported my academic research generously with annual unrestricted grants for 15 years. Also they prepared a number of intermediates for our research: several kilograms of the product of alkali fusion of Cleve's acid needed for the preparation of the intermediate **2** in our "pilot plant" production of equilenin, about 40 g of the product from the Bannister–Birch reduction of the 11-hydroxy substance **42**, and one kilo of 1-acetyl-1-methylcyclopropane for use in the Brady–Julia olefin synthesis. In the early 1960s, Sterling sent one of their process development specialists, Bill Wetterau, to Stanford to produce over a kilo of the key hydrochrysene intermediate **34** for use in our research.

Sterling's serious effort to commercialize our chloroquine synthesis was already described. Another involvement concerned a joint arrangement between Sterling and Chevron who carried out our JH synthesis on a large scale with the hope of developing it commercially under Stanford patents. As it turned out JH is not sufficiently stable in the field for use in insect control. However, Sterling, under the direction of Joe Collins, discovered that a late intermediate, the dione-ester, had antiviral activity, which led to the preparation of a number of analogs, some of which are quite promising. On October 28, 1981, Joe wrote to me as follows:

Our original premise in performing extensive antiviral testing on JH compounds was that the action of JH in regulating the expression of the adult gene in the insect may be applicable to the regulation of virus replication. Based on present information, it appears that this hypothesis may be valid and that we may have a handle on an extremely important and exciting new way of influencing the expression of extrachromosomal genetic information. If this is true, we may have a class of compounds with broad chemotherapeutic applications.

Denis Bailey (postdoc. 1961–1963), who is now Vice President for Drug Discovery at Sterling, recently advised me that this study still shows much promise, and two substances are undergoing clinical trials.

A joint project evolved from a study performed by Bob Christiansen on the cyanation of steroidal 4,6-dien-3-ones, which led to some highly strained bridged ring structures.[173] Another effort, carried out by Bob Clarke, involved the synthesis of some perhydroanthracene derivatives as possible steroid hormone analogs.[174] In the course of this study Bob was able to produce, among other isomers, a pair that differed only in that the central ring was chair or boat that was used in addition to our lactones (Scheme 9) for combustion calorimetry study.[175]

Another collaborative project developed from my association with Syd Archer, who was the person I spent the most time with during my visits to Sterling. Syd is an extraordinarily talented scientist who, while a graduate student at Penn State, became visible to the scientific community by his publication of the classical paper demonstrating that the Hoffmann rearrangement proceeds with retention of configuration.[176] He has since become one of the foremost medicinal chemists, winning the first Medicinal Chemistry Award, sponsored by the Medicinal Chemistry Division of the ACS. In 1968, he was made Associate Director and Divisional Vice President of the institute. Like me, Syd had started to lose his enthusiasm for Sterling after Suter's retirement in 1967. On the occasion of a dinner in Merle and Verle's honor, I presented the following limericks.

Tranquility (William and Barbara Allen Johnson)

The Institute leader from Chatham
Told his colleagues to get up and at 'em
From Alex's slurries
Came the treatment for worries
Trancopal was the answer verbatim.

Ode to Talwin (WSJ)

With the Demerol patent expiring
And Merle on the verge of retiring

Syd kept things serene
With a drug like Morphine
That the addicts found most uninspiring.

The Lathe Song (Barbara Allen Johnson)

There was a fine fellow named Merle
Who was turning a huge redwood burl
It grew thinner and flatter
But he said it's no matter
It will make a nice platter for Verle.

In 1973 Syd took early retirement to accept a professorship in chemistry at Rensselaer Polytechnic Institute, later becoming Dean of Science, but always keeping active in research. He is now doing the most important work of his career. I have always found discussing science with him especially stimulating, and we have enjoyed a long friendship with him and his wife Teddy.

Soon after our preliminary experiments on the synthesis of 16,17-dehydroprogesterone (Scheme 29) were completed, Syd raised the question of whether this methodology might be applicable to producing the 19-nor compound, which would be useful for making oral contraceptives. Information on model systems already in hand indicated that cyclization of the secondary alcohol **96** should give the 19-nor product **97** (Scheme 71). As it developed, Bob Clarke and Saul Daum at Sterling took on the project with great enthusiasm and very nicely reduced it to practice.[109] Scaleup was then started; however, the project was suddenly terminated. Bob Clarke disappointedly wrote to me on October 4, 1968: "It seems that the front office had the idea that we would be

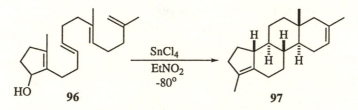

Scheme 71. Synthesis of a 19-nor compound.

finished with the whole project by now, having converted all intermediates to steroids for testing." This sort of thing occasionally makes industrial life difficult for serious scientists.

I was involved to some extent with the development of one of Sterling's commercial products: in the early days Ray Clinton was making some α-formyl ketones from steroidal unsaturated ketones, and I suggested that it might be interesting to convert these, with hydroxylamine, into the isoxazoles. This reaction was part of the chemistry we had used to produce the cyanoketone **12** in our equilenin synthesis, Scheme 4. These isoxazoles proved to be a new generation of pituitary-gonadotropic inhibitors, and one of them, Danazol, is on the market.

In 1947 Roger Adams, who was a consultant for DuPont for his whole professional life, was asked to recommend a young consultant for their Plastics Department, and he suggested my name. Consequently, I started to visit them when they were located in Newark, New Jersey, and later on when they moved to the Experimental Station in Wilmington, Delaware. It was during the early consulting days that I found myself on the same flight (back to Chicago) with Roger Adams. We managed to sit together and among other things Roger indicated that they were considerably concerned about and somewhat disappointed in their young nontenured assistant professor, E. J. Corey, because he seemed to be more interested in mechanism studies than in synthesis. Now I had already run into Corey on the beach near Salisbury, Massachusetts, where both of our families congregated around Labor Day. E. J. and I spent the better part of an afternoon talking and drawing formulas in the sand (Barbara heard E. J.'s aunt say "Those two boys have worms in their heads"). Well, I was tremendously impressed with this young man and I told Roger so with great conviction. Not very long after that episode, E. J. began to produce such sensational research results in synthesis that Roger became his staunchest supporter and he engineered Corey's promotion to full professor at age 27 to counter offers from other universities.

Consulting for DuPont was a highly educational experience for me, and I developed considerable interest in and a great respect for the chemistry of polymers. I remained with them until 1963 when I felt the need to curtail some of my outside activities, because of my involvement in administration at Stanford. I made a number of good friends at DuPont, the closest one being Art Anderson, a highly talented, creative scientist whom I knew from the very beginning in Newark. Eventually he became head of Exploratory Research in his department. I consulted occasionally for other departments such as Central Research and Organic Chemicals, and am co-inventor of one patent from these peripheral activities.

The DuPont crowd loved parties, in which I (sometimes with Barbara) often participated. When Sam Scott came to Wisconsin and later to Stanford on recruiting visits, he always spent time at our home, almost a part of the family. Sam, a Ph.D. organic chemist, may well have been the best recruiter of all time. He had a fantastic memory for people as well as their scientific accomplishments, and he was most welcome wherever he went. He had a wealth of information about major chemistry departments, and if you wanted to get the latest news on what professor was being lured from University X to University Y, you just asked Sam. After Sam died, we came to know George Boswell particularly well from his Stanford visits. It was through George's work, which he disclosed in a seminar at Stanford, that the incentive came for developing the very effective vinyl fluoride terminator for polyene cyclizations.[177] Altogether, 17 of my former pre- and postdoctoral associates have gone to DuPont, and they are widely scattered in different departments.

Carl Djerassi, the founder and president of the Zoecon Company, was responsible for Koji Nakanishi, Gilbert Stork, and me being consultants for several years. The small chemical research group, under the able guidance of Clive Henrick, hardly needed our help; hence, we met very infrequently. My major interaction was with the late John Siddall who worked independently of and served as an in-house consultant to Clive's group. John would often come over to my office to discuss chemistry, and as described on page 147, we became involved in a joint study of the chloro ketal Claisen reaction. John was a fine person and an extraordinarily talented scientist whom I found to be a most stimulating colleague. His untimely death due to leukemia was a tragic loss.

Other Industrial Contacts

One of my greatest benefactors is Pfizer Central Research in Groton, Connecticut, with whom I have enjoyed a close relationship largely because a number of my exceptionally good former students and postdoctoral assistants joined their research laboratories. Twelve such people, including President Barry Bloom and Executive Director of Medicinal Chemistry Chuck Harbert are at Groton, and three more are at Pfizer Ltd. in England, where Simon Campbell is Senior Director of Discovery Chemistry. I have visited the Groton labs several times, and over the years have enjoyed annual visits from Walt Moreland and now Chuck Harbert. Ever since 1974 the management has been making very generous unrestricted donations in support of my research program. Academic people find such funds of extraordinary value in many ways, particularly to back up commitments made to future collaborators who are hoping to be funded either on fellowships or on the PI's (principal investigator) pending federal grant proposals. Pfizer's total contribution of unrestricted funds to my program have now exceeded those that I have received from any other source.

Other companies with whom I have had especially good interaction, who have employed a number of my former collaborators, and who have made major contributions of unrestricted funds to my research effort include Hoffmann-LaRoche (Nutley), Upjohn, Searle, SmithKline Beecham, and Lilly.

Conclusion

No one questions the importance of organic synthesis, which is the backbone of the pharmaceutical industry as well as an integral part of many other scientific enterprises. I do not think that it is just a matter of coincidence that the most successful pharmaceutical companies have synthetic chemists in top managerial positions. Merck is a prime example having had leaders like Folkers, Hirschmann, Sarett, and Tishler. On the other hand, there is a general feeling entertained not only by lay scientists but even by chemists who have had little experience with the synthesis of moderately complex structures, that the field is simply a technology. Nothing could be further from the truth. Despite the great technological advances that have been made in the last 50 years, organic synthesis is still as much of an art as a science. Even with powerful spectroscopic and computer facilities as well as a large increase of the menu of methodological tools, we still face the same basic difficulties and uncertainties of 50 years ago because they are intrinsic to the field. The major difference is that today we can move much faster and address more difficult problems than in the past. The basic characteristics of the field of organic synthesis that have not changed and have hardly ever been appreciated by "outsiders" are considered here.

Planning versus Reduction to Practice

The present state of the art of synthesis of molecules of even modest complexity involves the same kinds of problems that were faced 50 years ago: all well-planned synthetic schemes almost invariably fail to give the envisaged results at one, or more often, several stages. Thus, a successful synthesis seldom follows and sometimes diverges dramatically from the original plan. It is this *fallibility phenomenon* that renders organic synthesis at least as much of a creative challenge at the execution stage as at the planning phase. History shows that the fallibility phenomenon has often been responsible for the generation of new synthetic methodology. This was the state of affairs in Robert Robinson's era, and it promises to persist for a long time to come. Scientists who have been seriously engaged in organic synthesis are so familiar with the fallibility phenomenon that they take it completely for granted. On the other hand, those scientists who have had little exposure to the field seldom have an appreciation of the phenomenon. I have found that my nonsynthetic colleagues hold to the belief that a well-conceived plan by an expert essentially assures success and that reducing the proposed scheme to practice is simply a routine technological matter. Except in the case of the most pedestrian synthetic objectives, nothing could be further from the truth.

The relatively recent availability of computer-assisted synthetic planning hardly alters the picture. This facility is potentially helpful at the planning stage, but advertisement of the tool as "programmed organic synthesis" is misleading and damaging to the cause by implying to the potential student, not to mention the general public, our governing bodies, and granting agencies, that synthesis is susceptible to automation. Acceptance of this attitude could be disastrous because I think we can make our most convincing justification of federal support for chemistry as a whole on the basis of synthesis alone, which is something that the general public can really appreciate.

Coping with the Fallibility Phenomenon

One might well wonder why it would require 55 years of work involving several hundred collaborators to accomplish what has been overviewed in the foregoing pages of this book. The answer is that dealing with the fallibility phenomenon, particularly as it applies to the development of new methodology, results in a high percentage of effort that is unsuccessful and is never published. I believe it was Emil Fischer who once said something to the effect that 90% of our efforts are of no great significance, but it is that 10% of real success that makes it all worthwhile.

In the game of reducing synthesis plans to practice, much effort can be expended in studying model systems that are relatively easy to produce. This can be a useful exercise; on the other hand it has serious disadvantages. A favorable result with a model series does not insure that success will be realized in the ultimate case. Of more serious concern is the fact that failure of a synthetic sequence in a model series spells doom for the plan, which consequently will almost never be tried in the originally scheduled environment where it might, in fact, have worked. Woodward eventually became very disenchanted with model systems and I have heard him say "there is only one good model system, namely the enantiomeric series of your target system".

The Human Element

People involved in synthesis must constantly rely on published experimental procedures. Those who are experienced in synthesis are generally pleasantly surprised when they are able to reproduce other people's experimental results. More often than not, the yield is lower than reported and sometimes the reaction does not work at all. Occasionally chemists cannot repeat their own experiments.[178] The reasons for this *irreproducibility phenomenon* are very complex, and the problem is exacerbated when the chemistry is being applied to a new substrate.[179] I started to learn about this problem while in graduate school when I wanted to carry out a Reformatsky reaction on a tetracyclic ketone in the chrysene series. The details of Bachmann's procedure for this reaction as applied to the keto ester **3**, to give >90% yield, were at hand. With my substrate the reaction stopped after about 20% completion and could not be restarted. During one of his visits to Harvard, I had the opportunity to discuss my problem with Bachmann who was extremely helpful. He started out by saying "The Reformatsky is one of those reactions you really have to make friends with." (I have since quoted his words on a number of occasions to students who were having trouble with tricky reactions.) He then quizzed me in detail about my experiments and, on learning that the product complex in my case precipitated as gummy material rather than a solid, he concluded that the recommended stirring of the reaction mixture was probably counter-productive in my example because it would help to coat the surface of the zinc with the gum so that the metal was no longer available as a reactant. On following his suggestion of omitting the stirring, the gum collected more on the walls of the flask and the reaction continued to about 80% completion.

People who have the expertise and intuition to resolve such ubiquitous problems are rare indeed. They enjoy what I call "the golden touch". I believe this gift is more subtle than just a matter of good training, which seemingly is a necessary but not sufficient condition. I have had several collaborators with the golden touch, but I shall mention only one, Chuck Harbert (postdoc. 1967–1969), because an example of his special talent is demonstrated by some of the chemistry already described.

In the late 1960s, we were using the Collins reagent routinely for oxidizing unsaturated alcohols like the dienol of Scheme 53 to aldehydes. The yields were generally about 70–80%. Chuck had occasion to carry out some of these oxidations and his yields were always >90%, such as 94% in the case shown in Scheme 53. Hoping to find the secret, I quizzed him in detail and came to the conclusion that, although he was a rapid worker, he also was more careful than others, paying more attention to using pure reagents. The substrate alcohol was evaporatively distilled, the pyridine and the chromium trioxide were rigorously dried, as was the dichloromethane solvent. (I believe the pyridine was dried by storage over barium oxide, a trick I learned at Kodak in conjunction with the preparation of cyanine dyes.) Relatively recently it has been shown that traces of water slow down these types of oxidations, and this may be part of the answer. Be that as it may, all of Chuck's experimental results were generally considerably better than the norm. He, like the other members of that rare group having the golden touch, appeared to have an intuitive feeling for the mechanics of a reaction and instinctively knew the right way to proceed, not unrelated to the well-known phenomenon of mechanical aptitude. Among my collaborators I have observed various degrees of this innate experimental aptitude.

An example of the other extreme is Tom Yarnell (Ph.D. 1975). Tom was an excellent and earnest scholar, who worked very hard at the bench developing what appeared to be very good technique, but he had extraordinary difficulty in duplicating procedures and getting new ones to work. He completed a very acceptable study for his thesis but it was a real *tour de force*, and he never seemed to enjoy his work. This may explain why he has since given up chemistry and obtained an M.D. As far as I could see, the motions Tom made at the bench were no different from those of skillful, productive experimentalists. This sort of thing is very mystifying.

Some Pros and Cons of Academic Life

The privilege of being free to work on research projects of your own choosing comes at a high price. Almost invariably it requires generating your own research funds from granting agencies, and this may require some compromise as to the area of research. Thus funds for supporting physical organic chemistry have for some time been seriously limited to meager NSF support—a great pity. People in the synthesis area are better off, because it is not difficult to make a case for application of almost any kind of synthesis methodology studies so that it will appeal to the NIH, which has a much larger budget than NSF.

Another problem that affects particularly the young faculty is the well-known pressure for publishing, which not only limits the research efforts to short range projects, but often creates a great deal of stress. Finally there is a disadvantage of carrying on a major research program in a university in the United States in that essentially all of the experimental work is performed by individuals who are students just learning the art. This is in contrast with industrial research where much of the work is performed by highly experienced experimentalists.

In view of these various problems, when I have been asked for advice, I have usually recommended that unless the student is absolutely determined to try to make a career as a research scholar in the academic environment, he would probably be better off in industry where the opportunity exists for performing research of the highest caliber without the aforementioned handicaps.

The case history of my former student Hans Wynberg illustrates the sort of determination that may be needed to surmount the hazards involved in an academic career. Hans' aim was to be a teacher at a good institution with a Ph.D. program. Such jobs were very scarce when he received his Ph.D. in 1952; so he spent a postdoctoral year at Minnesota. In 1953, the best job he could find was an assistant professorship at Grinnell College, a very good liberal arts institution, but without a graduate program. Working with undergraduates and his own hands, he started to make an impression on the scientific community by means of his publications, and in 1956 he accepted an associate professorship at Tulane. By working with graduate students, he began to become really visible. In 1960, after a year as a Fulbright Professor in the Netherlands (Leiden), he was invited to the University of Groningen, where he has made his name in science as he advanced to Full Professor and head of the Department of Organic Chemistry. The Wynberg story is a good lesson for young aspiring professors; they should not become discouraged if they are not successful in getting placed immediately in the university of their choice.

Dealing with students, however, has its own rewards, and this is one of the joys of academic life. Seeing young people get turned on by organic chemistry has been one of my greatest pleasures. The beginning organic course was always my favorite, for that is where so many of the students find for the first time that they really like chemistry. Occasionally there are a few students who become hooked, which I really enjoy vicariously as though I were back in my sophomore year at college. Also it is most fascinating to be closely involved with graduate students as they are learning to become research scholars, and to see what they do eventually with their careers. There are some real disappointments, such as the extraordinarily talented young prodigies who never really got their acts together, ending up as good but not outstanding scholars. I always wonder if I failed these students in some way. Gradually, I have learned not to get my hopes up prematurely, because it turns out that some exceptionally talented students have essentially reached their peaks early in their careers and do not improve. On the other hand, there are those scholars who have unending *capacity for improvement*, a quality that is often difficult to detect in the young student. The most distinguished scientists in the field of synthesis seem to have this characteristic, their most significant accomplishments generally occurring at middle age or later. Thus Woodward was starting to

Hans Wynberg, 1985.

reach his peak in the field at the age of 40 when some of his best efforts, namely the synthesis of strychnine and reserpine, were disclosed. Stork appears to be in a constant state of peaking, some of his best efforts being his latest, such as the intramolecular Diels–Alder approach to corticoids and his free-radical annulation methodology. Similarly Don Cram's best work surely is his latest. He has also written his autobiography, *From Design to Discovery* in the Profiles, Pathways, and Dreams series.

Among my own students who now are reaching the later years of their professional careers, Ralph Hirschmann and Bob Ireland stand out as two examples of chemists who have extraordinary capacity for improvement. Among Ralph's most significant contributions was his development of solution peptide synthesis methodology to the point where it is competitive with the Merrifield method as demonstrated by application to the first (simultaneously with Merrifield's) synthesis of an enzyme (i.e., ribonuclease). This work was announced when Ralph was 57 years old. (Bruce Merrifield has published his autobiography, *Life During a Golden Age of Peptide Chemistry*, within this Profiles, Pathways, and Dreams series.)

Bob Ireland was 47 when his seminal paper on the now well-known Ireland enolate Claisen rearrangement was published. He was 54 at the time his publication appeared on the synthesis of the complex ionophore antibiotic, lasalocid, which was achieved by methodology that rivals the best efforts of the masters in the field. One of the most gratifying aspects of being involved with teaching graduate students is to watch some of them develop into truly distinguished scholars.

Some Final Observations

I submit that the synthesis of complex molecules is a form of art like painting or architecture. Both areas require an immense amount of experience way beyond what can be learned from books or listening to lectures by experts. This experience in the field requires much time and effort, which may account for the fact that the great masters are apt to do their best work relatively late in their careers. This view is shared by an artist friend of ours,

Ralph Hirschmann during his early years at the Merck Research Laboratories in Rahway, NJ.   (Photo courtesy of R. Hirschmann.)

Earl Pierce, who is a painter, teacher (Barbara is a student of his), and art historian.

Organic synthesis may be very special in this respect: those involved in other areas like physical and theoretical chemistry seem to mature faster. This may account for the fact that organic synthetic chemists have won the ACS Award in Pure Chemistry (which is given only to chemists who are not over 36 years old) only eight times out of a total of 55. Note that Woodward did not receive this award.

Synthesis is and always has been the breeding ground for new chemistry. I remember Saul Winstein saying that the studies of the structure and synthesis of steroids provides one of the largest sources of "new" organic reactions and the understanding of their mechanisms. The failure of existing art to lead to needed structural or configurational molecular arrays inspires discovery of new reactions, mechanistic concepts, and methodologies. Irrespective of the immense practical importance of synthesis, the scientific consequences are in themselves a sufficient reason for engaging in the attempted synthesis of complex molecular structures. It is fortunate for the case of the advancement of chemistry that there have been and probably always will be a number of talented individuals who cast their lot with synthesis. The people who perform the experiments play the most important role in a synthetic endeavor. Progress in this field requires, in addition to special expertise, much hard work and a great deal of patience.

I have been most fortunate to have had many collaborators with these attributes, a large number of whom have permanently fallen in love with the field as I did. I recently wrote to all of my former collaborators with the aim of obtaining updated information about their professional as well as personal lives. The yield of replies has been good and I only wish it were possible to include comments about each of the individuals; however, space does not permit. Instead I am giving some summarized information about my professional progeny in Table I. There would have been nothing to write about in this book had it not been for these people.

Table I. Summary Information of Johnson Collaborators

Number	Category
5	Bachelor
19	Masters
101	Ph.D.
215	Postdoctoral associates
16	Scholars on sabbatical
356	Total

Proportion Involved in Various Career Activities

33% Academic:	25% Teaching and Research
	7% Administration
	1% Other
62% Industrial:	39% Experimentalist and Group Leaders
	23% Managerial Positions
5% Other	

A Tribute to My Co-workers

Among several awards that I have received during my career, the Tetrahedron Prize (1991) pleased me particularly because it afforded me the opportunity to publish an article entitled "Fifty Years of Research: A Tribute to My Co-Workers".[180] Thus I have been able to publish therein a list of my co-workers including some information about them, such as their location, etc. A few who were too recent to make the aforementioned list are all post-doctoral associates; namely Boris A. Czeskis, Paul A. Fish, Garth S. Jones, Cheol H. Lee, Gregory R. Luedtke, and Dayananda Raja-paksa. The *Tetrahedron* article[180] also contains a fairly complete list of the publications generated by me and my co-workers, which is organized according to topic.

One of the highest honors I have ever received was the invitation by the Harvard Chemistry Department for me and some of my previous co-workers to participate in a one-day affair on October 18, 1993, called the "Johnson Symposium". This was to be the first of a series honoring "distinguished" educators who are Harvard alumni. I was particularly delighted to receive this invitation because my co-workers were clearly being honored at the same time. As it turned out, it was with great reluctance that I found it necessary, after obtaining three medical opinions, to decline for health reasons. As I mentioned in my lengthy letter to the Harvard faculty (via David Evans), one of the cardiologists I consulted, who was obviously immensely impressed by such an invitation *from Harvard,* said in all seriousness "if you do go, you may end up in a hospital, but it might be well worth it".

I digress briefly to comment on my physical condition. My father, his mother, my sister, and my brother were all rather frail individuals with poor physical stamina. I, on the other hand, took after my mother's family who were moderately sturdy specimens. Although I was not really good at sports, I enjoyed them immensely, particularly tennis, gymnastics, and hiking. My general health was quite good well into my seventies when I began to have circulatory problems, which by now find me with the artery of my left leg completely blocked, which precludes my walking for more than 100 yards and requires my resting with the leg raised, preferably at heart-level, a good deal of the time. Ordinarily a problem like mine would be resolved by by-pass surgery;

Some of the former Johnson co-workers who were hired by Pfizer, U.S.A. Seated, left to right: Bob Volkmann, Chuck Harbert, Barry Bloom, Mike Hendrick, and John Lowe. Standing, left to right: Jim Korst, Glenn Andrews, and Tom Schaaf. This photograph was presented to me by Chuck Harbert on the occasion of a dinner for me in Miami on September 10, 1989, at the ACS meeting where I received the Arthur C. Cope Award.

however, a thorough cardiac examination reveals that I have two faulty valves that among other problems cause considerable arrhythmia. Also I have considerable enlargement of the heart. The experts have concluded that although the heart is functioning fairly well (with the help of Vasotec medication), a major operation of any sort would be too high a risk for me to withstand. The alternative of having complicated cardiac surgery followed by leg surgery later on is considered very high risk at my age.

I have given up accepting invitations to give lectures because I would have to do this from a lounge chair. The traveling would also be a problem. Even attending the Johnson Symposia at Stanford is very difficult for me, and recently I have had to skip afternoon talks in order to be able to be on hand at the dinner. I have learned that in situations where I am listening to a lot of exciting new chemistry and am having discussions with old and new friends, my adrenaline level rises to a point where I feel great and am completely unaware that I am overtaxing my circulatory system. It is impossible for me to monitor and control my condition under these circumstances, and the only solution is to avoid them. Fortunately I can still get a lot of enjoyment out of chemistry under relaxed conditions at home and at the laboratory without ending up being wiped out.

I have been able to drive my car to work, walk up to my office, and conduct business on about a half-time basis. I spend about three hours at the office in the morning, mainly at my desk with my legs on a support. At noon I go home, have lunch in a lounge chair, and take a short nap. In the afternoon I return to work for about two hours. By 4:00 p.m. I am very tired, return home, and go to sleep in my bed. About 6:30 I am good for a lounge-chair dinner. On rare occasions, Barbara and I will go out to dinner or to friends' homes in which case I will, of course, be equipped with leg support. My office door is open to visitors and I am lucky to have quite a few. All-in-all I am getting along quite nicely.

Returning to the Harvard symposium, it was decided to go ahead with the event anyway. My former students, Paul Bartlett and Bob Ireland, were the major speakers as originally planned. In addition Konrad Bloch, Dave Evans, and Frank Westheimer gave readings from portions of this book (prior to publication) which I had selected. Soon after the symposium, Harvard sent me a large package containing a beautiful, framed testimonial signed by the members of the chemistry department. This is reproduced in Figure 4. Also included were two VCR tapes containing a complete recording of the symposium, which I have now had the pleasure of "attending".

I was truly proud of the performances given by Bob and Paul. Evidently the Harvard people were favorably impressed, as indicated by a letter written to me by E. J. Corey. E. J. said, in part:

[The Johnson Symposium] was a great event in our department's history, one of the most memorable and moving of symposia in the traditional seminar room (MB-23), which has over the years been the scene of so much remarkable science. I was deeply moved and excited by your wonderful message. It also had a profound impact on our students and postdoctoral fellows. Your students Bob Ireland and Paul Bartlett gave splendid lectures; you would have been very proud of them. Although we all missed your presence, we are thankful to you for allowing the symposium to go forward. We had hoped that some miracle would happen and that somehow you would make a dramatic last-minute appearance. Anyhow, the Johnson day events are now officially a shining chapter in the continuing (we hope) legend of Harvard chemistry. ... Please give my best wishes to Barbara, and keep up the great work.

From the tone of this letter, I am inclined to feel that this symposium, in part because of the unusual format, was more successful than it would have been if I had attended. Either way, this was a splendid tribute to all of my co-workers.

HARVARD UNIVERSITY

DEPARTMENT OF CHEMISTRY

12 Oxford Street
Cambridge, Massachusetts 02138
U.S.A.

Professor William S. Johnson   *October 18, 1993*
Ph.D. Harvard University, 1940

Dear Professor Johnson,

We the Faculty of the Harvard University Department of Chemistry express our great admiration for your contributions to the field of organic chemistry. On this day, we wish to celebrate your distinguished career as a scientist and educator with the William S. Johnson Symposium.

Your extraordinary contributions to the synthesis of complex natural products have enriched and expanded the capacity of chemical synthesis to address complex architecture. Nowhere is this more evident than in your efforts to tame the cation cyclization approach to the steroid nucleus. The high standards which you have established in your scholarly achievements have served as a benchmark of excellence for the scientific community.

We also wish to honor you as an educator who has mentored more than three hundred and fifty collaborators, who, as young scientists, passed through your laboratory into the academic and industrial communities to become leaders in their disciplines.

On this special occasion we extend our very best wishes to you.

Sincerely,

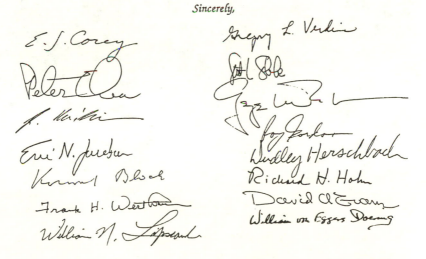

Figure 4. The "testimonial" sent by the Harvard University Chemistry Department staff to honor WSJ.

Bill and Barbara, on his birthday, February 24, 1995. Due to remodeling, Bill was forced to move to a temporary office.

"Instructor Johnson in his research laboratory," University of Wisconsin, ca. 1940.

William S. (Bill) Johnson
Scientist, Educator, Musician, Friend
Born in New Rochelle, NY
February 24, 1913
Passed Away at His Home
August 19, 1995

A Perspective on the Final Chemistry from Johnson's Group: An Epilogue

Paul A. Bartlett,[a] William R. Bartlett,[b] John D. Roberts,[c] and Gilbert Stork[d]

[a]University of California, Berkeley, CA 94720
[b]Fort Lewis College, Durango, CO 81301
[c]California Institute of Technology, Pasadena, CA 91125
[d]Columbia University, New York, NY 10027

Among the more spectacular applications of the evolving methodology of polycyclization are those which have been reported since Bill Johnson put down his pen for the last time while writing this Profile. These publications continue the themes that he and his co-workers pursued for some time, and whose origins he described above. As was characteristic of all of his research, there never seemed to be a dead end—no need to wind down an approach or wrap up a line of investigation. Indeed, like multiplying chain reactions, each inquiry spawned new ones, and the challenge was in setting priorities, rather than in trying to come up with new things to do. Among these last publications, we find communications of new synthetic methods in support of his polyene cyclization studies, further exploitation of the chiral acetal addition reactions,[181] continuing development of cyclization strategies and auxiliaries,[182] and their utilization in targeted synthesis of ever more complex structures.[183] These publications also include definitive reports of fluorine as a cation-stabilizing substituent in polyene cyclizations, expanding on the chemistry depicted in Schemes 67 and 70.

A head-to-head comparison of four traditional initiating moieties was undertaken, confirming that the tetramethylallyl and oxygen-stabilized cations that Bill's group had developed early on are among the most efficient. The smooth closure of each ring is favored by judicious placement of cation-stabilizing groups at sites where positive character is generated; this stabilization is provided by methyl groups where appropriate for the final structure, and by a fluorine atom at the position where the natural skeleton would lead the cyclization astray. The fluorine atom is readily lost in the Lewis acid-catalyzed pentacyclizations, where its elimination reduces ring strain, but in the synthesis of sophoradiol, it is retained in the pentacyclic product until elimination is required.

The acyclic initiators used in this study required that the cyclization process form five new rings rather than four (or fewer) as in all earlier cases. In the best cases studied, these non-enzymic *pentacyclizations* proceeded in ca. 50% yield, producing 7 stereocenters with the required stereochemistry of the oleanane nucleus! The epoxide **98a** (Scheme 72), while not producing **99** in as high yield (10% crystallized yield), was noteworthy in producing the pentacyclic ring system with all 8 stereocenters correctly established, including the 3β-hydroxyl group in the A ring. In addition, the 4,4-dimethyl substitution pattern found in many naturally occurring triterpenoids derived from oxidosqualene resulted. Sophoradiol (**104**) was prepared in only three steps from fluoropentacycle **103**, utilizing the tetramethylallyl alcohol **98d** as an "epoxide surrogate".

Closely related transformations of the *cis* and *trans* isomeric substrates provide another manifestation of the original promise of polyolefin cyclization: translation of olefin geometry, which is

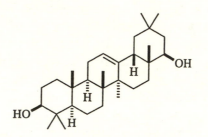

104

established relatively easily, into the ring-fusion stereochemistry of a polycyclic skeleton. Paul Fish and Garth Jones studied the cyclization of the chiral tetraenyne acetal *l-105* (Scheme 73), which makes an interesting comparison with the isomeric substrate **98c** (Scheme 72).[181]   In these reactions, the geometry of the last tri-

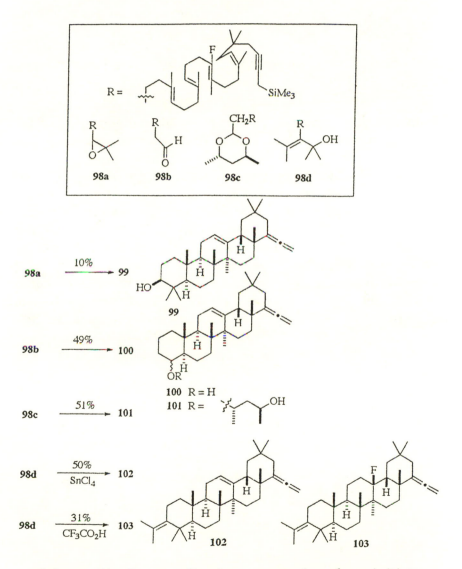

Scheme 72. Pentacyclizations comparing four initiator groups.

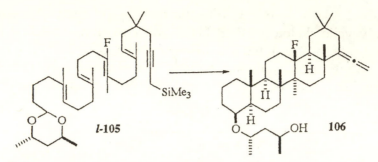

Scheme 73. Enantioselective formation of a pentacyclic compound.

substituted double bond is preserved in the configuration of the D–E ring junction, reflecting the stereospecific nature of the cyclizations.

What became Bill Johnson's final cyclization project aimed to combine the cyclization efficiency and latent A-ring functionality of the tetramethylallyl alcohol initiator with the chiral induction potential of the acetal initiator. This required development of a new chiral initiator. The allylic acetal initiator (see **107**) was designed by Bill, and model studies were performed by postdoctoral associate Boris Czeskis and Visiting Research Associate William R. (Ted) Bartlett at Stanford. Thus, substrate **107** (Scheme 74) was prepared and shown to cyclize efficiently (>85% GC yield) to give bicyclic alcohol 108 in one step with apparently greater than 90% ee according to Mosher ester analysis.[184]

The first target for the application of the new chiral acetal initiator was intended to be the dammarenediols, which had not, as yet in 1994, been the object of total synthesis. To evaluate the efficiency of the cylization process itself to form the dammarane

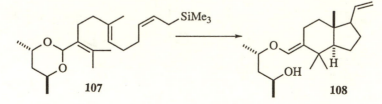

Scheme 74. Cyclization of the allylic acetal initiator.

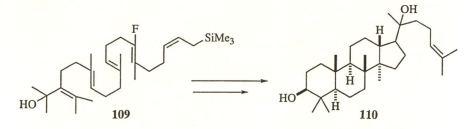

Scheme 75. Synthesis of the dammarenediols.

skeleton, an achiral allylic alcohol, substrate **109**, was prepared, having the well-studied tetramethylallyl alcohol initiator. Alcohol **109** was found to cyclize smoothly (61% yield) to give a tetracycle that was efficiently converted in several steps to the dammarenediols **110**.[185]

Sadly, Bill became terminally ill during the chiral acetal cyclization studies to make the dammarane ring system. As one of us (W.R.B.) remembers quite clearly, up to the time of his illness, Bill had driven from his home to his laboratories at Stanford every day to work at his desk for several hours. He would occasionally come into the labs to inquire about the lates results. If not, his office door was usually open and his gracious welcome was always assured. Brief progress reports would be listened to intently. After some thought, Bill would often summarize the essence of the results and provide an insightful suggestion to guide further work, drawing on his immense accumulated knowledge and intuition.

Upon Bill's passing away in August 1995, the postdoctoral associates in his research group continued their work, attempting to complete as much as possible of the ongoing projects. The enantioselective synthesis of the dammarenediols using the chiral acetal methodology could not be completed, however, and is currently being carried forward by one of us (W.R.B.) in the laboratories at Fort Lewis College. The Johnson laboratories at Stanford were closed in May 1998.

A Perspective on Bill Johnson and His Contributions to Chemistry

The overarching principles guiding the design of Bill Johnson's syntheses have been the efficient synthesis of the polyolefinic substrate itself, high yield and stereoselectivity in the cyclization process, and economy in converting the cyclization products to important, usually naturally occurring, polycyclic target compounds. Our present understanding of the process by which acyclic polyolefins form rings in high yield and with specific stereochemistry has evolved greatly from the first proof of the validity of the Stork–Eschenmoser hypothesis in 1964 (*see* Schemes 20 and 21). From its seemingly unpromising beginnings to its current, spectacular success required advances along many fronts: synthetic methods for assembling substrates stereoselectively, designs for initiators of appropriate reactivity that could be integrated into the final target, and strategies for enhancing and directing the polycyclization process itself.

An important legacy of this work is our knowledge of the mechanistic details of these complex transformations. For example, the requirement that each newly produced cationic center in the cyclization sequence be equal to or lower in energy than the previous one is illustrated by the effectiveness of cation-stabilizing auxiliaries such as the isobutenyl and fluoro substituents, and by the importance of terminating groups that disperse the positive charge quickly. The point-charge stabilization model of the enzymatic cyclization processes is another elegant outgrowth of the mechanistic understanding that emerged from this work. Ancillary discoveries (inventions, really), such as the ortho ester–Claisen rearrangement,[147] the modified-Julia method for construction of *E*-homoallylic bromides,[140] and the use of chiral acetals for asymmetric synthesis,[143,155] have in turn enlarged the scope of what can be done in the entire field of organic synthesis.

Bill Johnson's work didn't just alter the way we approach synthesis, it helped to change our view of what is even possible. As a result, an entire continent, stereospecific cationic cyclization, has been added to the world of organic chemistry. When he began his historic work on the polyene cyclization problem, no one seriously imagined that a complex structure like that of a natural steroid could, one day, be assembled with essentially complete regio- and stereospecificity; some actually stated so quite strongly! Although we are amused at how stunningly wrong this view turned out to be, we must not overlook the fact that when he started his career, the concept of controlling the stereochemical course of a sequence of organic reactions was unknown. The few syntheses of natural products that had been recorded (for example, camphor, cocaine, glucose, and hydroquinine) only served to emphasize the improbability of success in assembling more complex structures.

Organic chemists of the time approached syntheses by concentrating on the challenge of chemical connectivity; they simply ignored stereochemistry as an unreasonable handicap to the free exercise of their imagination. Although the ideas of regio- and stereoselectivity had yet to be articulated, Johnson made a conscious effort from the very start of his career to devise syntheses that implemented these concepts. Indeed, his first historically important contribution, a strategy for introducing an angular methyl group for steroid synthesis, established a pattern that would be one of the hallmarks of his research throughout his career: A synthetic method, born out of necessity for a larger goal, is developed into a general procedure with broad applicability from a well-reasoned mechanistic foundation. In addition to his conviction that mechanistic principles are crucial to the design of new synthetic methodology, we must also admire his long-term commitment to stay with a problem until it is solved.

Even within the context of modern synthetic strategies, Johnson's accomplishments in polyolefin cyclization stand out. The world of organic synthesis still largely depends on controlled methods for the construction of carbon–carbon bonds, of which there are not many. These methods have long been based on carbanion chemistry or electrocyclic processes and, more recently, on transition metal-mediated reactions. Ignoring conventional wisdom that carbocationic intermediates could not be controlled, Johnson foresaw their potential for contributing a major new methodology to organic synthesis. This vision, and Bill Johnson's persistence, account for the spectacular development of this methodology with which we are now familiar.

Seldom do we find major figures in organic synthesis whose work has such a strong central theme, and some find it surprising that Bill Johnson devoted so many years to that one area. However, accomplishments as significant as polyolefin cyclization do not come all at once in synthesis, and there is little doubt that the field would have lain fallow for a long time without Johnson's vision and his absolute dedication to achieving his goal. In other areas of organic synthesis to which many workers contributed, it is often difficult to untangle who was responsible for what. For cationic cyclization processes, however, Johnson's name is synonymous with the entire field. The reasons are clear: He had an unshakable conviction that, however unpromising early efforts at cationic cyclization might have been, no matter how skeptical the chemical community might be, treasures were likely to be found there. And, by the time others began to realize the potential of this chemistry, it seemed like his group was already doing everything that was worthwhile, with the result that the field was left entirely to Johnson.

In choosing to find a better, rather than the first, way to synthesize the compounds he was interested in, Bill Johnson chose a pathway different from that of many of his contemporaries. Although less glamorous, the greater long-term value of this approach was more in keeping with his personality and character. He never needed to show himself off as a "great man", although he was not lacking in competitive spirit; rather, he seemed to gain greater satisfaction with incremental progress. Perhaps it was this inner satisfaction that allowed Bill to remain productive to the end, despite the health problems that dogged him in his last years.

Bill Johnson's intellectual achievement was accompanied by the training of more than 350 scientific collaborators. He also contributed to the institutions he was associated with and spent many of his most productive years building a great chemistry department at Stanford. His participation in such activities as the Editorial Board of *Organic Syntheses* and the Chemistry Panel of the National Science Foundation also strengthened chemistry in other ways. Johnson's scientific integrity and unswerving expectation of scientific excellence—of himself and from his students—continue to be rewarded by his scientific legacy. We will not forget that he was also modest, gracious, and always considerate of others. Bill possessed a strong sense of humor; he was a truly enjoyable person. All who knew him will forever admire his chemical style and personal philosophies.

Acknowledgments

Individuals who have read and commented on various parts and versions of this book and whose help I am pleased to acknowledge with warmest thanks include: Robert Andrew, Ray Conrow, John Elliott, Mary Fieser, Barbara Johnson, Carolyn Southern, Jacquelyn Southern, and Gilbert Stork. I am particularly grateful to my editor, Jeffrey I. Seeman, who followed my efforts step-by-step and was enormously helpful.

I also wish to thank the various agencies listed in Table II who have provided generous financial support to my research efforts through my career.

| Table II. Major Sources of Research Support ||
Grant Funds	Unrestricted Gifts
National Institutes of Health	Hoffmann-LaRoche
National Science Foundation	Eli Lilly Company
Petroleum Research Fund, administrated	Pfizer
by the American Chemical Society	G. D. Searle Company
Wisconsin Alumni Research Foundation	SmithKline Beecham Institute
	Sterling-Winthrop Research
	Upjohn Company

Appendix I

Abbreviation	Possible Translation
B	Bill
QO	quick one
DM	diazomethane
W	write
U	you
OC	organic chemistry
T	theory
P	people, persons
SF	savoir faire
B	illegitimate child; alternatively, a disagreeable person
SU	Stanford University
OCD	organic chemical division
BCD	biochemical division
DD	Delores (Dodie) Dyer, RBW's secretary
PSARTFRS	Professor Sir Alexander R. Todd, FRS (Fellow of the Royal Society)

References

1. *Nobel Lectures Chemistry, 1922–1941;* Elsevier: Amsterdam, Netherlands, 1966. (b) Bachmann, W.; Cole, W.; Wilds, A. L. *J. Am. Chem. Soc.* **1940,** *62,* 824–841.
2. Johnson, W. S.; Mathews, F. J. *J. Am. Chem. Soc.* **1944,** *66,* 210–215.
3. Johnson, W. S.; Daub, G. H. In *Organic Reactions;* Adams, R., Ed.; John Wiley and Sons: New York, 1951; Chapter I, Vol. VI, p 19; *see also Org. Synth.* **1963,** *IV,* 132–135.
4. Newman, M. S.; Hart, R. T. *J. Am. Chem. Soc.* **1947,** *69,* 298–300.
5. Johnson, W. S.; Petersen, J. W.; Gutsche, C. D. *J. Am. Chem. Soc.* **1945,** *67,* 2274.
6. Johnson, W. S.; Petersen, J. W.; Gutsche, C. D. *J. Am. Chem. Soc.* **1947,** *69,* 2942–2955.
7. Johnson, W. S.; Christiansen, R. G.; Ireland, R. E. *J. Am. Chem. Soc.* **1957,** *79,* 1995–2005.
8. Johnson, W. S. *J. Am. Chem. Soc.* **1943,** *65,* 1317–1324.
9. Johnson, W. S. *J. Am. Chem. Soc.* **1944,** *66,* 215–217.
10. Johnson, W. S.; Banerjee, D. K.; Schneider, W. P.; Gutsche, C. D. *J. Am. Chem. Soc.* **1950,** *72,* 1426.
11. Johnson, W. S.; Banerjee, D. K.; Schneider, W. P.; Gutsche, C. D.; Shelberg, W. E.; Chin, L. J. *J. Am. Chem. Soc.* **1952,** *74,* 2832–2849.
12. Koebner, A.; Robinson, R. *J. Chem. Soc.* **1938,** 1994–1997.
13. Johnson, W. S.; Posvic, H. *J. Am. Chem. Soc.* **1945,** *67,* 504.
14. Johnson, W. S.; Posvic, H. *J. Am. Chem. Soc.* **1947,** *69,* 1361–1366.

15.   Birch, A. J.; Robinson, R. *J. Chem. Soc.* **1944,** 501–502.

16.   King, L. E.; Robinson, R. *J. Chem. Soc.* **1941,** 465–470.

17.   Birch, A. J.; Jaeger, R.; Robinson, R. *J. Chem. Soc.* **1945,** 582–586.

18.   Shorter, J. In the Centenary Tribute to Sir Robert Robinson; *Nat. Prod. Rep.* **1987,** *1(1),* 61–66.

19.   Johnson, W. S. In *Organic Reactions;* Adams, R., Ed.; John Wiley & Sons: New York, 1944; Vol. 2, Chapter 4.

20.   Johnson, W. S.; Woroch, E. L.; Buell, B. G. *J. Am. Chem. Soc.* **1949,** *71,* 1901–1905.

21.   Johnson, W. S.; Buell, B. G. *J. Am. Chem. Soc.* **1952,** *74,* 4513–4516.

22.   Johnson, W. S.; DeAcetis, W. *J. Am. Chem. Soc.* **1953,** *75,* 2766–2767.

23.   Johnson, W. S.; Schubert, E. N. *J. Am. Chem. Soc.* **1950,** *72,* 2187–2190.

24.   Winstein, S.; Goodman, L.; Boschan, R. *J. Am. Chem. Soc.* **1950,** *72,* 2311.

25.   McCasland, G. E.; Smith, D. A. *J. Am. Chem. Soc.* **1950,** *72,* 2190–2195.

26.   Barton, D. H. R. *Experientia* **1950,** *6,* 316–320.

27.   Johnson, W. S. *Experientia* **1951,** *7,* 315–319.

28.   Johnson, W. S.; Margrave, J. L.; Bauer, V. J.; Frisch, M. A.; Dreger, L. H.; Hubbard, W. N. *J. Am. Chem. Soc.* **1960,** *82,* 1255.

29.   Johnson, W. S.; Bauer, V. J.; Margrave, J. L.; Frisch, M. A.; Dreger, L. H.; Hubbard, W. N. *J. Am. Chem. Soc.,* **1961,** *83,* 606-614.

30.   Cornforth, J. W.; Robinson, R. *J. Chem. Soc.* **1949,** 1855–1865.

31.   Johnson, W. S.; Szmuszkovicz, J.; Rogier, E. R.; Hadler, H. I.; Wynberg, H. *J. Am. Chem. Soc.* **1956,** *78,* 6285–6289.

32.   Wilds, A. L.; Ralls, J. W.; Wildman, W. C.; McCaleb, K. E. *J. Am. Chem. Soc.* **1950,** *72,* 5794–5795.

33.   Johnson, W. S.; Bannister, B.; Bloom, B. M.; Kemp, A. D.; Pappo, R.; Rogier, E. R.; Szmuszkovicz, J. *J. Am. Chem. Soc.* **1953,** *75,* 2275–2276.

34.   Johnson, W. S.; Bannister, B.; Pappo, R. *J. Am. Chem. Soc.* **1956,** *78,* 6331–6339.

35. Cardwell, H. M. E.; Cornforth, J. W.; Duff, S. R.; Holter-
     mann, H.; Robinson, R. *Chem. Ind. (London)* **1951**, 389–390;
     J. Chem. Soc. **1953**, 361–384.

36. Woodward, R. B.; Sondheimer, F.; Taub, D.; Heusler, K.;
     MacLamore, W. M. *J. Am. Chem. Soc.* **1951**, *73*, 2403–2404; .
     J. Am. Chem. Soc. **1952**, *74*, 4223–4251.

37. Johnson, W. S.; Pappo, R.; Kemp, A. D. *J. Am. Chem. Soc.*
     1954, *76*, 3353–3354.

38. Johnson, W. S.; Kemp, A. D.; Pappo, R.; Ackerman, J.;
     Johns, W. F. *J. Am. Chem. Soc.* **1956**, *78*, 6312–6321.

39. Johnson, W. S.; Allen, D. S., Jr. *J. Am. Chem. Soc.* **1957**, *79*,
     1261.

40. Johnson, W. S.; Allen, D. S., Jr.; Hindersinn, R. R.; Sausen,
     G. N.; Pappo, R. *J. Am. Chem. Soc.* **1962**, *84*, 2181–2196.

41. Cole, J. E., Jr.; Johnson, W. S.; Robins, P. A.; Walker, J.
     Proc. Chem. Soc. **1958**, 114.

42. Cole, J. E., Jr.; Johnson, W. S.; Robins, P. A.; Walker, J. *J.
     Chem. Soc.* **1962**, *45*, 244–278.

43. Johnson, W. S.; Bannister, B.; Pappo, R.; Pike, J. E. *J. Am.
     Chem. Soc.* **1955**, *77*, 817–818.

44. Johnson, W. S.; Bannister, B.; Pappo, R.; Pike, J. E. *J. Am.
     Chem. Soc.* **1956**, *78*, 6354–6361.

45. Johnson, W. S.; Vredenburgh, W. A.; Pike, J. E. *J. Am.
     Chem. Soc.* **1960**, *82*, 3409–3415.

46. Keana, J. F. W.; Johnson, W. S. *Steroids* **1964**, *4*, 457–462.

47. Johnson, W. S.; Marshall, J. A.; Keana, J. F. W.; Franck, R.
     W.; Martin, D. G.; Bauer, V. J. *Tetrahedron* **1966**, Suppl. 8,
     Part II, 541–601.

48. Marshall, J. A.; Johnson, W. S. *J. Am. Chem. Soc.* **1962**, *84*,
     1485–1486.

49. Johnson, W. S.; Keana, J. F. W.; Marshall, J. A. *Tetrahedron
     Lett.* **1963**, 193–196.

50. Johnson, W. S.; Collins, J. C.; Pappo. R.; Rubin, M. B. *J. Am.
     Chem. Soc.* **1958**, *80*, 2585.

51. Johnson, W. S.; Collins, J. C., Jr.; Pappo. R.; Rubin, B. M.;
     Kropp, P. J.; Johns, W. F.; Pike, J. E.; Bartmann, W. *J. Am.
     Chem. Soc.* **1963**, *85*, 1409–1430.

52. Schiess, P. W.; Bailey, D. M.; Johnson, W. S. *Tetrahedron
     Lett.* **1963**, 549–553.

53. Johnson, W. S.; deJongh, H. A. P.; Coverdale, C. E.; Scott, J. W.; Burckhardt, U. *J. Am. Chem. Soc.* **1967**, *89*, 4523–4524.

54. Johnson, W. S.; Cohen, N.; Habicht, E. R., Jr.; Hamon, D. P. G.; Rizzi, G. P.; Faulkner, D. J. *Tetrahedron Lett.* **1968**, 2829–2833; *see also* Johnson, W. S.; Cox, J. M.; Graham, D. W.; Whitlock, H. W., Jr. *J. Am. Chem. Soc.* **1967**, *89*, 4524–4526.

55. Nakanishi, K.; Goto, T.; Ito, S.; Natori, S.; Nozoe, S. *Natural Products Chemistry;* Academic: Orlando, FL, 1974; Vol. 1, pp 496–497.

56. Barton, D. H. R.; Beaton, J. M. *J. Am. Chem. Soc.* **1961**, *83*, 4083–4089.

57. The term "partial synthesis" has been widely used by steroid chemists when referring to syntheses that are not "total" but use readily available (naturally derived) steroids as starting materials. The protocol for partial synthesis, therefore, involves elaboration of natural steroids.

58. Ireland, R. E. *Organic Synthesis;* Prentice Hall: Englewood Cliffs, NJ, 1969.

59. Woodward, R. B.; Bloch, K. *J. Am. Chem. Soc.* **1953**, *83*, 2023–2024, et seq. (Bloch).

60. Roberts, J. D. *On Thirty Years of Teaching and Research;* W. A. Benjamin: New York, 1970; pp 1-42 to 1-47.

61. Johnson, W. S.; Korst, J. J.; Clement, R. A.; Dutta, J. *J. Am. Chem. Soc.* **1960**, *82*, 614–622.

62. Roberts, J. D.; Regan, C. M.; Allen, I. *J. Am. Chem. Soc.* **1952**, *74*, 3679–3683.

63. Caserio, M. C.; Roberts, J. D.; Neeman, M.; Johnson, W. S. *J. Am. Chem. Soc.* **1958**, *80*, 2584.

64. Neeman, M.; Caserio, M. C.; Roberts, J. D.; Johnson, W. S. *Tetrahedron* **1959**, *6*, 36–47.

65. Johnson, W. S.; Neeman, M.; Birkeland, S. P. *Tetrahedron Lett.* **1960**, 1–5.

66. Johnson, W. S.; Neeman, M.; Birkeland, S. P.; Fedoruk, N. A. *J. Am. Chem. Soc.* **1962**, *84*, 989–992.

67. Barton, D. H. R.; Jeger, O.; Prelog, V.; Woodward, R. B. *Experientia* **1954**, *10*, 81–90.

68. Kupchan, S. M.; Johnson, W. S. *J. Am. Chem. Soc.* **1956**, *78*, 3864.

69. Kupchan, S. M.; Johnson, W. S.; Rajagopolan, S. *J. Am. Chem. Soc.* **1958**, *80*, 1769.

70. Kupchan, S. M.; Johnson, W. S.; Rajagopalan, S. *Tetrahedron* **1959,** *7,* 47–61.

71. Henbest, H. B.; Lovell, B. J. *Chem. Ind. (London)* **1956,** 278.

72. West, R.; Korst, J. J.; Johnson, W. S. *J. Org. Chem.* **1960,** *25,* 1976–1978.

73. The most important service performed by this particular committee was the establishment of the publication entitled *Directory of Graduate Research,* which evolved from a suggestion made by Arthur C. Cope, one of the members of the committee. The first issue of this biannual book appeared in 1953.

74. Franck, R. W.; Rizzi, G. P.; Johnson, W. S. *Steroids* **1964,** *4,* 463–481.

75. Nagata, W.; Terasawa, T.; Hirai, S.; Takeda, K. *Tetrahedron Lett.* **1960,** 27–33; *Tetrahedron,* **1961,** *13,* 295–307.

76. Kutney, J. P.; By, A.; Inaba, T.; Leong, S. Y. *Tetrahedron Lett.* **1965,** 2911–2918.

77. Kutney, J. P.; Cable, J.; Gladstone, W. A. F.; Hanssen, H. W.; Torupka, E. J.; Warnock,W. D. C. *J. Am. Chem. Soc.* **1968,** *90,* 5332–5334 *et sequ.*

78. Stork, G.; Burgstahler, A. W. *J. Am. Chem. Soc.* **1955,** *77,* 5068–5077; Stadler, P. A.; Eschenmoser, A.; Schinz, H.; Stork, G. *Helv. Chim. Acta* **1957,** *40,* 2191–2198; *see also* Burgstahler, A. W. Ph.D. Thesis, Harvard University, Cambridge, MA, August 1952; p 7.

79. Eschenmoser, A.; Ruzicka, L.; Jeger, O.; Arigoni, D. *Helv. Chim. Acta* **1955,** *38,* 1890–1904.

80. For a review of these studies *see* Eschenmoser, A.; Felix, D.; Gut, M.; Meier, J.; Stadler, P. In *Ciba Foundation Symposium on the Biosynthesis of Terpenes and Sterols;* Wolstenholme, G. E. W.; O'Conner, M., Eds.; J. and A. Churchill: London, 1959; pp 217–230.

81. van Tamelen, E. E.; Willett, J. D.; Clayton, R. B.; Lord, K. E. *J. Am. Chem. Soc.* **1966,** *88,* 4752–4754; Corey, E. J.; Russey, W. E.; Ortiz de Montellano, P. R. *J. Am. Chem. Soc.* **1966,** *88,* 4750–4751.

82. van Tamelen, E. E. *Acc. Chem. Res.* **1968,** *1,* 111–120; **1975,** *8,* 152–158.

83. van Tamelen, E. E.; Hwu, J. R. *J. Am. Chem. Soc.* **1983,** *105,* 2490–2491.

84.   Johnson, W. S. *Acc. Chem. Res.* **1968,** *1,* 1–8.

85.   Johnson, W. S. *Bioorg. Chem.* **1976,** *5,* 51–98.

86.   Bartlett, P. A. In *Asymmetric Synthesis;* Morrison, J. D., Ed.; Academic: Orlando, FL, 1984; Vol. 3, Chapter 4.

87.   Groen, M. B.; Zeelen, F. J. *Recueil Trav. Chim. Pays-Bas,* **1986,** *205,* 465-487.

88.   Bartlett, P. D. *Liebigs Ann. Chem.,* **1962,** *653,* 45-55; Bartlett, P. D.; Clossen, W. D.; Cogdell, T. J. *J. Am. Chem. Soc.,* **1965,** *87,* 1308–1314.

89.   (a) Johnson, W. S.; Bailey, D. M.; Owyang, R.; Bell, R. A.; Jaques, B.; Crandall, J. K. *J. Am. Chem. Soc.* **1964,** *86,* 1959–1966. (b) Johnson, W. S.; Crandall, J. K. *J. Am. Chem. Soc.* **1964,** *86,* 2085. (c) Johnson, W. S.; Owyang, R. *J. Am. Chem. Soc.* **1964,** *86,* 5593–5598. (d) Johnson, W. S.; Crandall, J. K. *J. Org. Chem.* **1965,** *30,* 1785–1790. (e) Johnson, W. S.; Kinnel, R. B. *J. Am. Chem. Soc.* **1966,** *88,* 3861–3862.

90.   Johnson, W. S. *Pure Appl. Chem.* **1963,** *7,* 317–334.

91.   Johnson, W. S.; van der Gen, A.; Swoboda, J. J. *J. Am. Chem. Soc.* **1967,** *89,* 170–172.

92.   van der Gen, A.; Wiedhaup, K.; Swoboda, J. J.; Dunathan, H. C.; Johnson, W. S. *J. Am. Chem. Soc.* **1973,** *95,* 2656–2663.

93.   Johnson, W. S.; Jensen, N. P.; Hooz, J. *J. Am. Chem. Soc.* **1966,** *88,* 3859–3860.

94.   Johnson, W. S.; Jensen, N. P.; Hooz, J.; Leopold, E. J. *J. Am. Chem. Soc.* **1968,** *90,* 5872–5881.

95.   Bartlett, P. A.; Johnson, W. S. *J. Am. Chem. Soc.* **1973,** *95,* 7501–7502.

96.   Bartlett, P. A.; Brauman, J. I.; Johnson, W. S.; Volkmann, R. A. *J. Am. Chem. Soc.* **1973,** *95,* 7502–7504.

97.   Abrams, G. D.; Bartlett, W. R.; Fung, V. A.; Johnson, W. S. *Bioorg. Chem.,* **1971,** *1,* 243-268.

98.   Johnson, W. S.; Semmelhack, M. F.; Sultanbawa, M. U. S.; Dolak, L. A. *J. Am. Chem. Soc.* **1968,** *90,* 2994–2996.

99.   Johnson, W. S.; Li, T.-t.; Harbert, C. A.; Bartlett, W. R.; Herrin, T. R.; Staskun, B.; Rich, D. H. *J. Am. Chem. Soc.* **1970,** *92,* 4461–4463.

100. Johnson, W. S.; Gravestock, M. B.; Parry, R. J.; Myers, R. F.; Bryson, T. A.; Miles, D. H. *J. Am. Chem. Soc.* **1971,** *93,* 4330–4332.

101.   Gravestock, M. B.; Johnson, W. S.; McCarry, B. E.; Parry, R. J.; Ratcliffe, B. E. *J. Am. Chem. Soc.* **1978,** *100,* 4274–4282.

102.   Markezich, R. L.; Willy, W. E.; McCarry, B. E.; Johnson, W. S. *J. Am. Chem. Soc.* **1973,** *95,* 4414–4416.

103.   McCarry, B. E.; Markezich, R. L.; Johnson, W. S. *J. Am. Chem. Soc.* **1973,** *95,* 4416–4417.

104.   Johnson, W. S.; McCarry, B. E.; Markezich, R. L.; Boots, S. G. *J. Am. Chem. Soc.* **1980,** *102,* 352–359.

105.   Morton, D. R.; Gravestock, M. B.; Parry, R. J.; Johnson, W. S. *J. Am. Chem. Soc.* **1973,** *95,* 4417–4418.

106.   Morton, D. R.; Johnson, W. S. *J. Am. Chem. Soc.* **1973,** *95,* 4419–4420.

107.   Johnson, W. S.; Schaaf, T. K. *Chem. Commun.* **1969,** 611.

108.   (a) Volkmann, R. A.; Andrews, Glenn C.; Johnson, W. S. *J. Am. Chem. Soc.* **1975,** *97,* 4777–4779. (b) Johnson, W. S.; Shenvi, A. B.; Boots, S. G. *Tetrahedron* **1982,** *38,* 1397–1404.

109.   Daum, S. J.; Clarke, R. L.; Archer, S.; Johnson, W. S. *Proc. Natl. Acad. Sci. U.S.A.* **1969,** *62,* 333–336.

110.   Johnson, W. S.; Huffman, W. F.; Boots, S. G. *Recl. Trav. Chim. Pays-Bas* **1979,** *98,* 125–126.

111.   (a) Johnson, W. S.; Berner, D.; Dumas, D. J.; Nederlof, P. J. R.; Welch, J. *J. Am. Chem. Soc.* **1982,** *104,* 3508–3510; (b) Johnson, W. S.; Dumas, D. J.; Berner, D. *J. Am. Chem. Soc.* **1982,** *104,* 3510–3511.

112.   Marsham, P.; Widdowson, D. A.; Sutherland, J. K. *J. Chem. Soc., Perkin Trans. 1* **1974,** 238–241.

113.   (a) Peters, J. A. M.; Posthumus, T. A. P.; van Vliet, N. P.; Zeelen, F. J.; Johnson, W. S. *J. Org. Chem.* **1980,** *45,* 2208–2214. (b) Seeman, J. I. *Chem. Rev.* **1983,** *83,* 83–134.

114.   Johnson, W. S.; Daub, G. W.; Lyle, T. A.; Niwa, M. *J. Am. Chem. Soc.* **1980,** *102,* 7800–7802.

115.   Johnson, W. S.; Newton, C.; Lindell, S. D. *Tetrahedron Lett.* **1986,** *27,* 6027–6030.

116.   Johnson, W. S.; Hughes, L. R.; Kloek, J. A.; Niem, T.; Shenvi, A. *J. Am. Chem. Soc.* **1979,** *101,* 1279–1281.

117.   Johnson, W. S.; Hughes, L. R.; Carlson, J. L. *J. Am. Chem. Soc.* **1979,** *101,* 1281–1282.

118.   Schmid, R.; Huesmann, P. L.; Johnson, W. S. *J. Am. Chem. Soc.* **1980,** *102,* 5122–5123.

119. Johnson, W. S.; Yarnell, T. M.; Myers, R. F.; Morton, D. R.; Boots, S. G. *J. Org. Chem.* **1980,** *45,* 1254–1259.

120. Hughes, L. R.; Schmid, R.; Johnson, W. S. *Bioorg. Chem.* **1979,** *8,* 513–518.

121. Johnson, W. S.; Lyle, T. A.; Daub, G. W. *J. Org. Chem.* **1982,** *47,* 161–163.

122. Johnson, W. S.; Bunes, L. A. *J. Am. Chem. Soc.* **1976,** *98,* 5597–5602.

123. Garst, M. E.; Cheung, Y.-F.; Johnson, W. S. *J. Am. Chem. Soc.* **1979,** *101,* 4404–4406.

124. van Tamelen, E. E.; Willet, J.; Schwartz, M.; Nadeau, R. *J. Am. Chem. Soc.* **1966,** *88,* 5937–5938.

125. van Tamelen, E. E.; Leiden, T. M. *J. Am. Chem. Soc.* **1982,** *104,* 2061–2062.

126. (a) Johnson, W. S.; Wiedhaup, K.; Brady, S. F.; Olson, G. L. *J. Am. Chem. Soc.* **1968,** *90,* 5277–5279. (b) Johnson, W. S.; Wiedhaup, K.; Brady, S. F.; Olson, G. L. *J. Am. Chem. Soc.* **1974,** *96,* 3979–3984.

127. Johnson, W. S.; Chen, Y.-Q.; Kellogg, M. S. *J. Am. Chem. Soc.* **1983,** *105,* 6653–6656.

128. Johnson, W. S.; Telfer, S. J.; Cheng, S.; Schubert, U. *J. Am. Chem. Soc.* **1987,** *109,* 2517–2518.

129. Johnson, W. S.; Escher, S.; Metcalf, B. W. *J. Am. Chem. Soc.* **1976,** *98,* 1039–1041.

130. Johnson, W. S.; Brinkmeyer, R. S.; Kapoor, V. M.; Yarnell, T. M. *J. Am. Chem. Soc.* **1977,** *99,* 8341–8343.

131. Brinkmeyer, R. S.; Kapoor, V. M. *J. Am. Chem. Soc.* **1977,** *99,* 8339–8341.

132. Johnson, W. S.; Fagundo, C.; Ravelo, A. G., unpublished observations.

133. Speziale, A. J.; Stephens, J. A.; Thomson, Q. E. *J. Am. Chem. Soc.* **1954,** *76,* 5011–5013; Barkley, L. B.; Farrar, M. W.; Knowles, W. S.; Raffelson, H.; Thompson, Q. E. *J. Am. Chem. Soc.* **1954,** *76,* 5014–5016.

134. Johnson, W. S.; Lindell, S. D.; Steele, J. *J. Am. Chem. Soc.* **1987,** *109,* 5852–5853.

135. *See* footnote 22 of reference 128 and footnote 15 of reference 134.

136.   Compare the stabilization of the cationic transition state by aspartate-52 in the classical Phillips mechanism for lysozyme. For a documented involvement of an external point charge in the binding site of bovine rhodopsin *see* Honig, B.; Dinur, U.; Nakanishi, K.; Balogh-Nair, V.; Gawinowicz, M. A.; Arnaboldi, M.; Motto, M. G. *J. Am. Chem. Soc.* **1979,** *101,* 7084–7086.

137.   van Tamelen, E. E.; Anderson, R. J. *J. Am. Chem. Soc.* **1972,** *94,* 8225–8228.

138.   Cornforth, J. W. In *Ciba Foundation Symposium on the Biosynthesis of Terpenes and Sterols;* Wolstenholme, G. E. W.; O'Connor, M., Eds.; Little, Brown and Co.: Boston, MA, 1958; pp 299–300.

139.   Ourisson, G. In *Natural Products and Biological Activities;* Imura, H.; Goto, T.; Murachi, T.; Nakajima, T., Eds.; University of Tokyo: Tokyo, Japan, 1984; p 55.

140.   Brady, S. F.; Ilton, M. A.; Johnson, W. S. *J. Am. Chem. Soc.* **1968,** *90,* 2882–2889.

141.   Julia, M.; Julia, S.; Guégan, R. *Bull. Soc. Chim. Fr.* **1960,** 1072–1079; Julia, M.; Julia, S.; Tchen, S.-Y. *Bull. Soc. Chim. Fr.* **1961,** 1849–1853.

142.   Roberts, J. D.; Mazur, R. H. *J. Am. Chem. Soc.* **1951,** *73,* 2509–2520.

143.   (a) Johnson, W. S.; Harbert, C. A.; Stipanovic, R. D. *J. Am. Chem. Soc.* **1968,** *90,* 5279–5280. (b) Johnson, W. S.; Harbert C. A.; Ratcliffe, B. E.; Stipanovic, R. D. *J. Am. Chem. Soc.* **1976,** *98,* 6188–6193.

144.   Parker, K. A.; Johnson, W. S. *Tetrahedron Lett.* **1969,** 1329–1332.

145.   Johnson, W. S.; Li, T.-t.; Faulkner, D. J.; Campbell, S. F. *J. Am. Chem. Soc.* **1968,** *90,* 6225–6226.

146.   Johnson, W. S.; Campbell, S. F.; Krishnakumaran, A.; Meyer, A. S. *Proc. Natl. Acad. Sci. U.S.A.* **1969,** *62,* 1005–1009.

147.   Johnson, W. S.; Werthemann, L.; Bartlett, W. R.; Brocksom, T. J.; Li, T.-t.; Faulkner, D. J.; Petersen, M. R. *J. Am. Chem. Soc.* **1970,** *92,* 741–743.

148.   Faulkner, D. J.; Petersen, M. R. *Tetrahedron Lett.* **1969,** 3243.

149. Miles, D. H.; Loew, P.; Johnson, W. S.; Kluge, A. F.; Meinwald, J. *Tetrahedron Lett.* **1972,** 3019–3022.

150. Johnson, W. S.; Brocksom, T. J.; Loew, P.; Rich, D. H.; Werthemann, L.; Arnold, R. A.; Li, T.-t.; Faulkner, D. J. *J. Am. Chem. Soc.* **1970,** *92*, 4463–4464.

151. Prestwich, G. D.; Labovitz, J. N. *J. Am. Chem. Soc.* **1974,** *96*, 7103–7105.

152. (a) Loew, P.; Siddall, J. B.; Spain, V. L.; Werthemann, L. *Proc. Natl. Acad. Sci. U.S.A.* **1970,** *62*, 1462–1464. (b) Werthemann, L.; Johnson, W. S. *Proc. Natl. Acad. Sci. U.S.A.* **1970,** *67*, 1810–1813.

153. Loew, P.; Johnson, W. S. *J. Am. Chem. Soc.* **1971,** *93*, 3765–3766.

154. Werthermann, L.; Johnson, W. S. *Proc. Natl. Acad. Sci. U.S.A.* **1970,** *67*, 1465–1467. (b) Werthemann, L.; Johnson, W. S. *Proc. Natl. Acad. Sci. U.S.A.* **1970,** *67*, 1810–1813.

155. Johnson, W. S.; Elliott, J. D.; Hanson, G. *J. Am. Chem. Soc.* **1984,** *106*, 1138–1139.

156. Bartlett, P. A.; Johnson, W. S.; Elliott, J. D. *J. Am. Chem. Soc.* **1983,** *105*, 2088–2089.

157. Johnson, W. S.; Crackett, P. H.; Elliott, J. D.; Jagodzinski, J. J.; Lindell, S. D.; (in part) Natarajan, S. *Tetrahedron Lett.* **1984,** *25*, 3951–3954.

158. Johnson, W. S.; Chan, M. F. *J. Org. Chem.* **1985,** *50*, 2598–2600.

159. Elliott, J. D.; Choi, V. M. F.; Johnson, W. S. *J. Org. Chem.* **1983,** *48*, 2294–2295.

160. Choi, V. M. F.; Elliott, J. D.; Johnson, W. S. *Tetrahedron Lett.* **1984,** *25*, 591–594.

161. Johnson, W. S.; Elliott, R.; Elliott, J. D. *J. Am. Chem. Soc.* **1983,** *105*, 2904–2905.

162. Lindell, S. D.; Elliott, J. D.; Johnson, W. S. *Tetrahedron Lett.* **1984,** *25*, 3947–3950.

163. Johnson, W. S.; Edington, C.; Elliott, J. D.; Silverman, I. R. *J. Am. Chem. Soc.* **1984,** *106*, 7588–7591.

164. Silverman, I. R.; Edington, C.; Elliott, J. D.; Johnson, W. S. *J. Org. Chem.* **1987,** *52*, 180–183.

165. Elliott, J. D.; Steele, J.; Johnson, W. S. *Tetrahedron Lett.* **1985,** *26*, 2535–2538.

166. Johnson, W. S.; Fletcher, V. R.; Chenera, B.; Bartlett, W. R.; Tham, F. S.; Kullnig, R. K., *J. Am. Chem. Soc.* **1993,** *115,* 497–504.

167. Johnson, W. S.; Buchanan, R. A.; Bartlett, W. R.; Tham, F. S.; Kullnig, R. K. *J. Am. Chem. Soc.* **1993,** *115,* 504–515.

168. Johnson, W. S.; Plummer, M. S.; Reddy, S. P.; Bartlett, W. R. *J. Am. Chem. Soc.* **1993,** *115,* 515–521.

169. Fish, P. V.; Pulla Reddy, S.; Lee, C. H.; Johnson, W. S. *Tetrahedron Lett.* **1992,** *33,* 8001–8004.

170. Groen, M. B.; van Vliet, N. P.; Zeelen, F. J. *Symposium Papers of the 11th IUPAC International Symposium on the Chemistry of Natural Products*; International Union of Pure and Applied Chemistry: London, 1978; Vol. 3, pp 29–31, *et seq.*

171. Dijkink, J.; Speckamp. W. N. *Tetrahedron* **1978,** *34,* 173–178, *et seq.*

172. Corvers, A.; van Mil, J. H.; Sap, M. M. E.; Buck, H. M. *Recl. Trav. Chim. Pays-Bas* **1977,** *96,* 18–22 *et seq.*

173. Christiansen, R. G.; Johnson, W. S. *Steroids* **1963,** *1,* 620–627.

174. Clarke, R. L.; Johnson, W. S. *J. Am. Chem. Soc.* **1959,** *81,* 5706–5710.

175. Margrave, J. L.; Frisch, M. A.; Bautista, R. G.; Clarke, R. L.; Johnson, W. S. *J. Am. Chem. Soc.* **1963,** *85,* 546–548.

176. Archer, S. *J. Am. Chem. Soc.* **1940,** *62,* 1872–1874.

177. *See* footnote 7 of reference 114.

178. For example, *see* the letter of Murphy, H. W. *Chem. Eng. News* **1987,** *June 29,* 4.

179. It was mainly because of the irreproducibility phenomenon that the publication *Organic Syntheses* was conceived by Roger Adams who was the editor of the first volume in 1920. The protocol of *Organic Syntheses* is to publish selected (submitted) procedures that are then carefully checked by the Board of Editors. Even with this special care, these procedures are not always reproducible.

180. Johnson, W. S., *Tetrahedron* **1991,** *41,* xi–l. (This article, entitled *Fifty Years of Research, A Tribute to My Co-Workers,* was written by the author on the occasion of his receipt of the Tetrahedron Prize in 1991.)

181. Fish, P. V.; Johnson, W. S.; Jones, G. S.; Tham, F. S.; Kullnig, R. K. *J. Org. Chem.* **1994,** *59,* 6150–6152.
182. (a) Fish, P. V.; Johnson, W. S. *J. Org. Chem.* **1994,** *59,* 2324–2335. (b) Fish, P. V.; Johnson, W. S. *Tetrahedron Lett.* **1994,** *35,* 1469–1472.
183. Fish, P. V.; Sudhakar, A. R.; Johnson, W. S. *Tetrahedron Lett.* **1993,** *34,* 7849–7852.
184. Johnson, W. S.; Czeskis, B.; Bartlett, W. R., Stanford University, unpublished results.
185. Johnson, W. S.; Czeskis, B.; Bartlett, W. R.; Lemoine, R.; Gautier, A.; Luedtke, G.; Leopold, E.; Bancroft, K.J., Stanford University, unpublished results.

Index